LIVING LIGHT

Dragon-fish pursuing luminescent squid. Reproduced by kindness of Wm. Beebe.

Painting by E. Bostelmann

LIVING LIGHT

BY

E. NEWTON HARVEY

*"These blazes . . .
giving more light than heat"*
HAMLET, ACT I, SCENE 3

(facsimile of 1940 edition)

HAFNER PUBLISHING COMPANY, INC.
New York   London
1965

Originally published 1940

reprint 1965, with corrections

———

Printed and Published by

HAFNER PUBLISHING COMPANY, INC.

31 East 10th Street

New York, N.Y. 10003

Library of Congress Card Catalogue Number 65-28700

© Copyright, 1940, Princeton University Press

All rights reserved. No part of this book may be reproduced without permission of the publisher.

CONTENTS

INTRODUCTION   xi

I. COLD LIGHT   3

Sensitivity of eye. Luminescence widespread. "Shining fish and flesh." "Burning of the sea." Luminous animals. False reports of luminescence. Secondary luminescence and bacterial infection. Symbiosis. Reflection phenomena. Purple arcs of eye. "Flashing of flowers." St. Elmo's fire. Luminous rain and snow. *Ignis fatuus.* Luminous sweat and urine.

II. LIGHT-PRODUCING ORGANISMS   23

A general account of the behavior and structure of luminous groups including: bacteria, fungi, radiolaria, flagellates, sponges, jellyfish, hydroids, siphonophores, pennatulids, ctenophores, nemerteans, bryozoans, marine worms, earthworms, crustacea, myriopods, insects, molluscs, squid, ophiuroids, balanoglossids, ascidians, fishes.

III. TYPES OF LUMINESCENCE   88

Theory of light emission. Black body radiation. Incandescence and luminescence. Candoluminescence. Pyroluminescence. Thermoluminescence. Phosphorescence and fluorescence. Electroluminescence. Sonoluminescence. Galvanoluminescence. Triboluminescence or Piezoluminescence. Crystalloluminescence and Lyoluminescence. Chemiluminescence or Oxyluminescence. Bioluminescence or Organoluminescence.

IV. CHEMISTRY OF LIGHT PRODUCTION   122

Luminescence and water. Luminescence and oxygen. Carbon dioxide production. Photogen. Luciferin and luciferase. Photophelein. Specificity of photogenic substances. Properties of photogenic substances. Proluciferin. Oxidized luciferin. Luciferesceine. Models of bioluminescence. Kinetics of bioluminescence. Mechanism of bioluminescence. Evolution of living light.

V. PHYSIOLOGY OF LIGHT PRODUCTION   160

Stimulation. Narcosis. Light. Radium rays. Temperature. Salts. Hydrogen ion concentration. Respiratory poisons. Respiratory accelerators.

CONTENTS

VI. PHYSICAL NATURE OF ANIMAL LIGHT   194
    Measurement of light. Intensity of animal light. Quality of animal light. Efficiency. Invisible radiation. Mitogenetic rays.

BIBLIOGRAPHY   227

INDEX   259

ILLUSTRATIONS

Frontispiece—Dragon-fish pursuing luminescent squid. *facing title page*
Courtesy of Wm. Beebe, from a painting by Else Bostelmann.

Fig. 1 Man's face photographed by fluorescent light excited by ultraviolet without visible rays. *page* 257
Courtesy of Akademische Gesellschaft, Leipzig.

Fig. 2 Two dead herring photographed by the light of luminous bacteria growing on them. 258
After Pratje, from a photograph of Professor Rosen.

Fig. 3 Spruce slab infiltrated with the mycelium of a fungus, often responsible for "shining wood." 258
Courtesy of Henry Holt and Company.

Fig. 4 Large luminous animals of the sea, two shrimp and a jellyfish. 259
After Macartney, 1810.

Fig. 5 Comb-jellies (*Pleurobrachia*). 260
Courtesy of Chicago University Press, from a photograph by F. Schensky.

Fig. 6 Microscopic organisms responsible for "burning of the sea." 261
Above. Dinoflagellates after Ehrenberg, 1834.
Below. *Noctiluca*, after Quatrefages, 1850.

Fig. 7 A fish of the Banda Sea, *Photoblepharon palpebratus*. 262
Courtesy of Ulric Dahlgren, after a drawing by Bruce Horsfall.

Fig. 8 Luminous bacterial colonies and individual bacteria. 263
After Molish and Kishitani.

Fig. 9 Luminous bacteria in a round dish, photographed by their own light (right), and by ordinary light (left). 264
Photographed by Goro, Black Star, by courtesy of "Life."

Fig. 10 A sand-flea such as is often afflicted by a luminous bacterial malady. 265
Courtesy of Chicago University Press and American Nature Association.

Fig. 11 The mycelium of *Armillaria mellea*, a luminous fungus found growing on wooden supports of coal mines. 265
Courtesy of Henry Holt and Company.

Fig. 12 Luminous fishes of the Banda Islands. Two *Photoblepharon* (above) and two *Anomalops* (below), from an original photograph. 266

Fig. 13 A cross section of the luminous organ of *Photoblepharon*. 267
After Steche.

ILLUSTRATIONS

Fig. 14 Structures in the luminous cells of *Pyrosoma*, believed to be symbiotic bacteria. 267
After Pierantoni.
Fig. 15 The fish, *Monocentris japonica*. 267
After Yasaki.
Fig. 16 The fish, *Physiculus japonicus*. 268
After Kishitani.
Fig. 17 The mushroom, *Cliocybe illudens*, in which the fruiting body is luminous. 269
Courtesy of Henry Holt and Company.
Fig. 18 The luminous mycelium of the coffee leaf spot fungus, *Omphalia flavida*, growing in a Petri dish. 269
Courtesy of A. H. R. Buller.
Fig. 19 The luminous leaf spot fungus, *Omphalia flavida*, on a coffee leaf. 270
Courtesy of A. H. R. Buller.
Fig. 20 The stag-horn fungus, *Xylaria hypoxylon*, reported by some to be luminous. 270
Courtesy of A. H. R. Buller.
Fig. 21 A modern figure of *Noctiluca*, by day (left) and at night (right). 271
After Pratje.
Fig. 22 The oldest figures of *Noctiluca*. 271
After Dicquemare, 1775; Slabber, 1778, and Macartney, 1810.
Fig. 23 Dinoflagellates (*Peridinium Bahamense*), responsible for the luminescence of "Fire Lake" in the Bahamas. 271
After Plate.
Fig. 24 Various types of luminous unarmored dinoflagellates (mostly *Gymnodinium*). 272
After Kofoid and Swezy.
Fig. 25 A radiolarian, *Thalassicolla nucleata*. 273
After Huth.
Fig. 26 The armored dinoflagellate, *Gonyaulax polyhedra*. 273
After Buhigas.
Fig. 27 Luminous hydroids by day and at night. 274
After Panceri.
Fig. 28 The sea pansy, *Renilla*, a pennatulid. 274
After Parker.
Fig. 29 Sea pens, luminous colonies of animals. 275
Courtesy of Ulric Dahlgren, from a drawing by Bruce Horsfall.
Fig. 30 *Mnemiopsis leidyi*, a common ctenophore. 276
Courtesy of University of Chicago Press.
Fig. 31 *Pelagia noctiluca*, a luminous jellyfish. 276
After Steuer.
Fig. 32 Ctenophores or comb-jellies, by day and at night. 277
After Panceri.

ILLUSTRATIONS

Fig. 33 *Chaetopterus* by day, after Trojan; at night, after Panceri; *Mesochaetopterus* at night, after Fujiwara. 278
Fig. 34 *Chaetopterus* pulled out of its tube by an eel. 279
Courtesy of Ulric Dahlgren, from a drawing by Bruce Horsfall.
Fig. 35 Sections of the luminous gland of *Cypridina*. 280
After Yatsu and Dahlgren.
Fig. 36 Transverse section of the luminous layer of *Chaetopterus*. 280
After Dahlgren.
Fig. 37 A luminous cirratulid worm. After Panceri.
Tomopteris rolasi, after von Greef. 281
Fig. 38 The luminous polynoëd worm, *Acholoë astericola*. 281
After Kutschera.
Fig. 39 A luminous copepod (*Pleuromma abdominale*). 281
After Giesbrecht.
Fig. 40 A polynoid worm attacked by a crab. 282
Courtesy of Ulric Dahlgren, from a drawing by Bruce Horsfall.
Fig. 41 *Cypridina hilgendorfii*. Original photographs. 283
Fig. 42 Sections of photophores of shrimp, *Sergestes prehensilis*.
After Terao. *Acanthephyra debilis*. After Kemp. 284
Fig. 43 A deep-sea shrimp, *Acanthephyra purpurea*, secreting from its luminous gland during a battle with the fish, *Photostomias guernei*. 285
Courtesy of the National Geographic Society, after a painting by E. J. Geske.
Fig. 44 Some luminous insects. A *Collembolan*. After Henneguy. Non-luminous adult and luminous larva of a fungus gnat, *Ceratoplanus testaceus*. After Stammer. Larva, pupa and adult of firefly, *Photuris pennsylvanica*. After Williams. 286
Fig. 45 Transverse section of the larval light organ of a Japanese glowworm, *Pyrocoelia rufa*. 287
After Okada.
Fig. 46 The light organ at the ends of the Malphigian tubules of the New Zealand glowworm. 287
After Wheeler and Williams.
Fig. 47 Transverse section of the light organ of an adult firefly, *Photinus consanguineus*. 287
After Williams.
Fig. 48 The West Indian firefly, *Pyrophorus noctilucus*. After Mangold. The related nonluminous "death watch." A newly hatched larva of *Pyrophorus* with its luminous organ. After Dubois. 288
Fig. 49 The female beetle *Phengodes* by day and at night. Original photograph. 289
Fig. 50 A common firefly on a leaf and a photograph of its light at night. 289
Photo of C. Clarke. By courtesy of the Chicago University Press.

ILLUSTRATIONS

Fig. 51 The boring marine mussel, *Pholas dactylus*, by day and by night. 290
After Panceri.

Fig. 52 The pelagic shell-less snail, *Phillirrhoë bucephala*, by day and at night. 291
After Panceri.

Fig. 53 A large luminous deep-sea squid catching lantern-fish. 292
Courtesy of Wm. Beebe, after a painting by Else Bostelmann.

Fig. 54 A deep-sea squid, *Lycoteuthis* (*Thaumatolampas*) *diadema*. 293
After Chun.

Fig. 55 *Sepia officinalis* and *Rondeletia minor*. 293
After Meissner.

Fig. 56 Snake-stars or brittle stars. 293
Courtesy of the Chicago University Press.

Fig. 57 The Japanese firefly squid, *Watasenia scintillans*, by day and at night. 294
After Sasaki.

Fig. 58 Transverse sections of the light organs of two squid, *Abraliopsis* and *Calliteuthis*. 78
After Chun.

Fig. 59 Original photograph of three *Pyrosoma* and a section of the colony. 295
After Panceri.

Fig. 60 The deep-sea fish, *Chauliodus sloanei*, attacking smaller forms. 296
Courtesy of Wm. Beebe, from a painting by Else Bostelmann.

Fig. 61 The hatchet fish, *Argyropelecus*. 297
Courtesy of Ulric Dahlgren, from a drawing by Bruce Horsfall.

Fig. 62 A deep-sea fish and deep-sea squid. 298
Courtesy of Ulric Dahlgren, from a drawing by Bruce Horsfall.

Fig. 63 Section of light organs of fish, *Stomias* and *Cyclothone*. 299
After Brauer.

Fig. 64 Pale round-mouths, *Cyclothone braueri*. 300
Courtesy of Wm. Beebe.

Fig. 65 *Echiostoma tanneri* and *Linophryne arborifera*. The double row of light organs of *Cyclothone*. 301
Courtesy of Wm. Beebe.

Fig. 66 *Idiacanthus fasciola* chasing shrimp. 302
Courtesy of Wm. Beebe from a painting by Else Bostelmann.

Fig. 67 *Grammatostomias dentatus*. 303
Courtesy of Wm. Beebe from a painting by Else Bostelmann.

Fig. 68 Electromagnetic waves. 304
After Henshaw.

Fig. 69 A string galvanometer record of the light of luminous bacteria after oxygen lack. Original record. 304

ILLUSTRATIONS

Fig. 70 Spectral distribution of black body radiation at different temperatures. 92
Courtesy of John Wiley and Sons.

Fig. 71 Spectral energy distribution of some phosphorescences. 101
After Coblentz and Hughes.

Fig. 72 Bacteria (*Bacillus anthracis*) photographed in fluorescent light; dark field ultraviolet and transmitted ultraviolet light. 305
After Barnard and Welch.

Fig. 73 Spectral energy distribution of the chemiluminescence of dimethyldiacridylium nitrate. 120
After Eymers and van Schouwenburg.

Fig. 74 Relation of the luminescence intensity of a fresh water bacterium (*Vibrio phosphorescens*) to oxygen pressure. 124
After Shapiro.

Fig. 75 Relation of the total oxygen consumption of a marine luminous bacterium to oxygen pressure. 125
After Shoup.

Fig. 76 The decay curve of luminescence of *Cypridina* luciferin and luciferase. 150
After Stevens.

Fig. 77 String galvanometer record of the decay of light when *Cypridina* luciferin and luciferase are mixed. 306
After Harvey and Snell.

Fig. 78 String galvanometer records of two flashes of the firefly, *Photuris pennsylvanica*. 307
After Snell.

Fig. 79 String galvanometer record of the light of the elaterid bettle, *Pyrophorus*. Original record. 307

Fig. 80 Relation between the light intensity and duration of flashes of a firefly at different oxygen tensions. 169
After Snell.

Fig. 81 Relation between oxygen consumption of a fresh-water luminous bacterium, *Vibrio phosphorescens*, and temperature. 181
After Root.

Fig. 82 Relation between luminescence intensity of a fresh-water luminous bacterium, *Vibrio phosphorescens*, and temperature. 182
After Root.

Fig. 83 The effect of diluted and of concentrated sea water on the luminescence and oxygen consumption of a salt-water luminous bacterium, *Achromobacter fischeri*. 186
After Johnson and Harvey.

Fig. 84 Visibility of radiation curves for an average light adapted and dark adapted eye. 195
Courtesy of McGraw Hill Book Co.

ILLUSTRATIONS

Fig. 85 Spectral energy distribution of the firefly and *Cypridina* light. 203
After Coblentz and Hughes.

Fig. 86 Spectral energy curves for *Photobacterium phosphoreum*. 205
After Eymers and van Schouwenburg.

Fig. 87 Luminous efficiency of the carbon incandescent lamp and the firefly. 208
After Ives and Coblentz.

Fig. 88 Luminous efficiency of the luminous bacterium, *Photobacterium phosphoreum*. Redrawn by the author from a curve of Eymers and van Schouwenburg. 210

INTRODUCTION

TWENTY-FIVE years have passed since the author first started investigation of living light and twenty years since the publication of *The Nature of Animal Light*. From 1920 on, much progress has been made in our knowledge of light emission by living things—bioluminescence. "Living light" includes this new work but is much more than a second edition of the older book. New chapters dealing with the morphology and physiology of light production in various groups of luminous animals have been introduced and other aspects completely rewritten to include the recent experiments. Special attention is devoted to various types of non-living luminescence. The historical approach has been adopted and each chapter will be found complete in itself, presenting the subject matter of interest to biologist, chemist or physicist.

Much of the literature on animal light is made up of reports that this or that animal is luminous or of records of an especially brilliant phosphorescence of the sea. Among those who have inquired more fully into the nature and mechanism of light production may be mentioned Anderson, Beijerinck, Dahlgren, Dubois, Ehrenberg, Heller, Johnson, Kanda, Kishitani, Krukenberg, Mangold, McDermott, Okada, Panceri, Phipson, Pierantoni, Pratje, Quatrefages, Spallanzani, Tilesius, Trojan, van Schouwenburg and Zirpolo. It is surprising that Darwin scarcely mentions animal light in his "Origin of Species."

To these names must be added the great scientists in other fields, men whose interest was aroused by animal light but whose energy was expended in other directions. Boyle, Newton, Franklin, Priestley, Reamur, A. von Humboldt, Des-

INTRODUCTION

saignes and Becquerel have all investigated luminous forms. Sir Humphry Davy studied the effect of hydrogen gas on the luminescence of a glowworm and Michael Faraday, during his Italian journey, endeavored to determine if firefly light was dependent on the life of the animal.

Indeed, living light has interest for all with scientific curiosity in the fields of biology, chemistry, physics and illuminating engineering. It is noteworthy that the last two decades have witnessed a rapidly increasing use for all kinds of cold light—luminous paints, fluorescent screens for television and new types of glow lamp and fluorescent lamp for illumination. The incandescent filament is destined to become a thing of the past, along with gas and the Welsbach mantle. This revolution in lighting has again returned us from incandescence to luminescence, the method of the firefly, a pioneer in the science of illumination.

It is always a pleasure to acknowledge the assistance of others. I pay special tribute to my students and associates whose investigation has done so much to advance the knowledge of animal light and whose publications are listed in a special section of the bibliography; to my colleagues, Drs. Frank H. Johnson, Aurin M. Chase, L. A. Turner and K. W. Cooper for reading certain chapters, and to my wife, Dr. Ethel B. Harvey, for careful reading and criticism of the whole book; finally to Mr. Charles Butt, my Research Assistant for the past ten years, whose aid in carrying out experiments has been invaluable.

Because of space limitations it has not been possible to print a list of known luminous species but the synonymy of the names of genera containing luminous forms has been checked by specialists in various fields: Drs. Waldo L. Schmitt, W. L. Tressler, W. M. Tattersall, Elizabeth Deichmann, Wm. W. Diehl and Miss E. K. Cash, J. P. Moore, Paul Bartsch, R. S.

INTRODUCTION

Bassler, H. L. Clark, C. A. Kofoid, J. T. Nichols, W. G. Van Name, C. B. Wilson, R. V. Chamberlin and E. A. Chapin. I am deeply grateful for the willing assistance of these men.

Professor Ulric Dahlgren has very kindly given me access to his reprint collection and allowed me to use some of the special studies of luminous animals in their native haunts, drawn by Bruce Horsfall. Dr. William Beebe, with whom I have spent some delightful hours deep-sea fishing in Bermuda, has loaned me the original paintings by Mrs. Else Bostelmann, of luminous fish, one of which is used as a frontispiece.

LIVING LIGHT

"These blazes . . . giving more light than heat"
—*Hamlet, Act I, Scene 3*

CHAPTER I

COLD LIGHT

A TROPIC noon with its dazzling brilliance and a moonlit night are two extremes in which the eye is equally at home. One is hot, the world of incandescence, the other cool, a realm of luminescence; the one a million times brighter than the other. To perceive this enormous range of brightness a change in the retina from light to dark adaptation must occur. The eye must act and does act like a balance that can easily weigh both a ton and a gram.

It is the ton in light to which we are accustomed. Man is so fundamentally diurnal in habit that he is seldom aware of the universal occurrence of faint luminescences—light emissions with no heat, produced under the most unexpected circumstances. How many have seen the light that appears when tire-tape or surgeon's adhesive tape is stripped from a roll? Although faint, this phenomenon is easily observed by the dark adapted eye. Lumps of sugar will luminesce when rubbed together. Diamonds and many other crystals when fractured emit a flash of light. Quartz pebbles glow when struck; sheets of mica when split. Even a rubber band may luminesce when snapped.

Many solutions of organic substances will produce light if ozone is bubbled through them or if treated with strong oxidizing agents. Some of these chemical lights are indeed brilliant when we consider that no heat accompanies them. The luminescence of phosphorous, a slow oxidation, has been observed since its discovery by Brandt in 1669.

Early also (1602) came knowledge of the stone, which, when heated with charcoal, gave *lapis solaris* or Bologna phosphorus, an impure barium sulphide which "imbibed light" and gave it off in the dark. John Canton's phosphorus, impure calcium sulphide, was later (1768) prepared by heating oyster shells and sulphur. Hulme spoke of it as the "light magnet of Canton," because of its power of attracting and absorbing light. These phosphori, which can store up light and re-emit it afterwards, aroused great interest in the seventeenth century. A large number are now known and form the basis of various luminous paints. This persistence of luminescence after illumination is spoken of as phosphorescence although the word is also used as a general term synonymous with luminescence. Greeks called the morning star "Phosphorus," Romans called it "Lucifer." Both names play a large part in the nomenclature of this subject.

Phosphori emit light while they are being illuminated, as well as afterwards, although it is not ordinarily detectable because of the high intensity of the illuminating light. A large group of compounds also luminesce while they are being illuminated although they emit no light afterwards. One method to demonstrate this is to use an "exciting" light such as ultraviolet, invisible to the eye. Then the luminescence is easily observed and is called fluorescence. Many minerals and almost all organic compounds fluoresce in ultraviolet light. Skin, nails, bone, teeth and proteins are especially brilliant (Fig. 1). An ultraviolet world would be one, not of total darkness, but of ghostly lights, with here and there a colored patch of some brilliant fluorescent substance to relieve the dull monotony of dimmest gray.

Cathode rays, X-rays, and various elementary particles in addition to light, can also excite fluorescence. This is the light of television, fluorescence of material on the kinescope to reproduce the image. Another well known example is the luminous paint on watch dials. The luminescence results from bombardment of a specially prepared zinc sulphide by alpha particles emitted from a trace of added radium salt. The impact of each individual alpha particle can easily be seen if the material is examined with a lens. Crookes used this method in 1903 to count the alpha particles from radium and called the apparatus a spinthariscope.

Then there are the various types of luminescence in gases, excited electrically, in vacuum tubes of one kind or another, neon, argon, nitrogen, mercury or sodium vapor lamps, which have converted our principal thoroughfares into a riot of color.

Laboratory workers may have seen the greenish glow emitted where mercury vapor condenses in a vacuum pump or the luminescence over the surface of an aluminum anode in electrolytic rectifiers. Even sound waves passing through a liquid produce light if they are intense enough. This extraordinary form of luminescence can be observed in pure water where sound waves cause cavitation with rapid separation of gas as bubbles from solution.

Thus a world without a sun and without fire would by no means be a dark world. Its creatures would have eyes adapted to faint lights, a retina comparable to and probably better than our own peripheral retina, which is more sensitive to light than our fovea, the central spot of greatest visual acuity. Such conditions of permanent darkness occur in the depths of the ocean and it is just here that we find living creatures to have evolved with enormous eyes and the most complicated pattern of luminous organs. Indeed the organs themselves are veritable lanterns with lenses, reflectors, pigment screens and photogenic material, which can be turned on and off "at will."

In addition to the deep-sea fauna living in utter darkness, the world of cold light is filled with innumerable surface and terrestrial forms, all examples of bioluminescence. Fireflies and glowworms are familiar, but the light of damp wood and dead fish and flesh is less common. Aristotle mentions the light of dead fish (Fig. 2) and flesh and both Aristotle and Pliny that of damp wood (Fig. 3). Robert Boyle in 1667 made many experiments to show that the light from these sources, as well as that of the glowworm, is dependent upon a plentiful supply of air and drew an interesting comparison between the light of shining wood and that of a glowing coal.

Boyle's observations on luminous meat were made in 1672 when a servant found a brightly luminous neck of veal in his cellar. "Notwithstanding the great Number of lucid Parts not the least degree of Stench was perceivable to infer any Putrefaction." Thinking that atmospheric conditions had something to do with the luminescence, Boyle carefully noted that "*The Wind as far as we could observe it, was then at South-west, and blustering enough; the Air by the sealed Thermoscope, appeared hot for the season, the Moon was past its last Quarter; the Mercury in the Barometer stood at 29 3/16 inches.*" He had no means of finding out the true cause of the light and early views of its nature were indeed fantastic. Even as late as 1800, Hulme concludes from his experiments on phosphorescent fish that the light is a "constituent principle of marine fishes" and the "first that escapes after the death of the fish." Although Baker (1742) had suggested that fish luminescence might be due to animacules, it was much later that Michaelis (1830) and Ehrenberg (1834) decided that the light of flesh must be the result of some living thing and that Heller (1854) gave the name *Sarcina noctiluca* to the suspected organism. In 1875 Pflüger showed that nutrient media could be inoculated with small amounts of luminous fish and that these would increase in size, like bacterial colonies, and we now know that the light of all dead fish and flesh is due to luminous bacteria growing on them.

COLD LIGHT

In the early part of the nineteenth century it was surmised that the light of damp wood was connected with fungus growth because of a similarity in smell. In 1854 Heller actually saw minute living strands, which he called *Rhizomorpha noctiluca*, as the source of the light. We now know that all phosphorescence of wood comes from strands or mycelia of various kinds of fungi and that sometimes the fruiting body of the fungus also produces light.

The phosphorescence or "burning of the sea," which is described by so many of the older explorers, also results entirely from living organisms, both microscopic and macroscopic. The latter are mostly jellyfish (Fig. 4) or comb jellies (Fig. 5) and give rise to the larger, more brilliant flashes of light often seen in the wake or about the sides of a steamer at night. The former are various species of flagellates (Fig. 6), such as *Noctiluca* (just visible to the naked eye) which collect at the surface of the sea and often increase in such numbers that the water is colored pink by day and shines like a sheet of fire when disturbed at night.

Father Bourzes noted these colored organisms without knowing what they were. After carefully describing the phosphorescence of Indian Seas, observed in 1704, he wrote as follows to the Royal Society of London: "*If one takes some Water out of the Sea, and stirs it never so little with his Hand in the dark, he may see in it an infinite number of bright Particles. Or if one dips a piece of Linnen in Sea Water, and twists or wrings it in a dark Place, he shall see the same thing; and if he does so, though it be half dry, yet it will produce abundance of bright Sparks. When one of the Sparkles is once formed, it remains a long time; and if it fix upon any thing that is solid, as for instance, on the side or edge of a Vessel, it will continue shining for some Hours together. It is not always that this Light appears, tho' the Sea be in great Motion; nor does*

LIVING LIGHT

it always happen when the Ship sails fastest: Neither is it the simple beating of the Waves against one another that produces this Brightness, as far as I could perceive: But I have observ'd that the beating of the Waves against the Shore, has sometimes produced it in great plenty; and on the Coast of Brazil the Shore was one Night so very bright, that it appeared as if it had been all on Fire.

"The Production of this Light depends very much on the Quality of the Water. . . . And I have often observed, that when the Wake of the Ship was brightest, the Water was more fat and glutinous; and Linnen moisten'd with it produced a great deal of Light, if it were stir'd or mov'd briskly. Besides, in sailing over some Places of the Sea, we find a Matter or Substance of different Colours, sometimes red, sometimes yellow. In looking at it, one would think it was Saw-dust: Our Sailors say it is the Spawn or Seed of Whales. What it is, is not certain; but when we draw up Water in passing over these Places, it is always viscous and glutinous. Our Mariners also say, That there are a great many Heaps or Banks of this Spawn in the North; and that sometimes in the Night they appear all over of a bright Light, without being put in Motion by any Vessel or Fish passing by them. But to confirm farther what I say, viz. That the Water, the more glutinous it is, the more it is disposed to become luminous, I shall add one particular which I saw my self. One Day we took in our Ship a Fish, which some thought was a Boneta. The inside of the Mouth of the Fish appeared in the Night like a burning Coal; so that without any other Light, I could read by it the same Characters that I read by the Light in the Wake of the Ship. It's Mouth being full of a viscous Humour, we rubbed a piece of Wood with it, which immediately became all over luminous; but as soon as the Moisture was dried up, the Light was extinguish'd.

COLD LIGHT

"*I leave it to be examined whether all these particulars can be explained by the system of such as assert, that the principle of this light consists in the motion of a subtle matter, or globules, caused by a violent agitation of different kinds of salts.*"

The light of the sea was a mysterious phenomenon to the older observers. Boyle's[1] ideas were expressed as follows: "*And when I remember, how many questions I have asked navigators about the luminousness of the sea; and how in some places the sea is wont to shine in the night as far as the eye can reach; at other times and places, only when the waves dash against the vessel, or the oars strike and cleave the water; how some seas shine often, and others have not been observed to shine; how in some places the sea has been taken notice of, to shine when such and such winds blow, whereas in other seas, the observation holds not; and in the same tract of sea, within a narrow compass, one part of the water will be luminous, whilst the other shines not at all: when I say, I remember how many of these odd phaenomena belonging to those great masses of liquor, I have been told of by very credible eye-witnesses (whose narratives to me you may elsewhere meet with) I am tempted to suspect, that some cosmical law or custom of the terrestrial globe, or, at least, of the planetary vortex, may have a considerable agency in the production of these effects.*"

Most of the older explanations of sea light were purely physical, ignition of sea-salt (Papin, 1647) or rubbing of salt molecules giving sparks like flint (Cartesius, 1648). Benjamin Franklin has been widely quoted as believing the phosphorescence to be electrical in origin but he changed his opinion as the result of two simple experiments performed in 1740. He writes: "*In my former paper on this subject, wrote first in 1747, enlarged and sent to England in 1749, I considered the sea as*

[1] From Dr. Birch's first edition of Boyle's Collected Works (1744), Vol. III, p. 91.

LIVING LIGHT

the grand source of lightening, imagining its luminous appearance to be owing to electric fire, produc'd by friction between the particles of water and those of salt. Living far from the sea, I had then no opportunity of making experiments on the sea water, and so embraced this opinion too hastily.

"For in 1750 and 1751, being occasionally on the sea coast, I found, by experiments, that sea water in a bottle, tho' at first it would by agitation appear luminous, yet in a few hours it lost that virtue; hence, and from this, that I could not by agitating a solution of sea salt in water produce any light, I first began to doubt of my former hypothesis, and to suspect that the luminous appearance in sea water, must be owing to some other principles."

In 1853, a letter from J. B. Esq. (James Bowdoin) of Boston to B. F., published in Franklin's *Letters and Papers on Philosophical Subjects* (1759), pointed out that the light in sea water could be removed by filtering through a cloth and that the "said appearance might be caused by a great number of little animals, floating on the surface of the sea, which, on being disturbed, might, by expanding their finns, or otherwise moving themselves, expose such a part of their bodies as exhibits a luminous appearance, somewhat in the manner of a glow-worm, or fire-fly. . . ." Franklin replied as follows: ". . . It is indeed very possible, that an extremely small animalcule, too small to be visible even by the best glasses, may yet give a visible light. I remember to have taken notice, in a drop of kennel water, magnified by the solar microscope to the bigness of a cart-wheel, there were numbers of visible animalcules of various sizes swimming about; but I was sure there were likewise some which I could not see, even with that magnifier; for the wake they made in swimming to and fro was very visible, though the body that made it was not so. Now, if I could see the wake of an invisible animalcule, I imagine I might much more easily see its light if it were of the luminous kind. For how small is the extent of a ship's wake, compared with that of the light of her lantern."

Certain knowledge that some of the phosphorescence of the sea is due to organisms came in the middle of the eighteenth century when small luminous worms and crustaceans were noted and *Noctiluca*, the organism frequently responsible for sea light, was first recognized as a luminous animal by Baker (1753).

Macartney, speaking before the Royal Society in 1810, sums up the situation as follows: "*Many writers have ascribed the light of the sea to other causes than luminous animals. Martin supposed it to be occasioned by putrefaction, Silberschlag believed it to be phosphoric; Professor J. Mayer conjectured that the surface of the sea imbibed light, which it afterwards discharged. Bajon and Gentil thought the light of the sea was electric, because it was excited by friction. . . . I shall not trespass on the time of the Society to refute the above speculations; their authors have left them unsupported by either arguments or experiments, and they are inconsistent with all ascertained facts upon the subject. The remarkable property of emitting light during life is only met amongst animals of the four last classes of modern naturalists, viz., mollusca, insects, worms, and zoöphytes.*" Macartney recognized the true cause of the light, although he had little idea of the vast number of marine forms which are luminous and omits entirely any reference to the fishes, many of which produce a light of their own when living, apart from any bacterial infection.

A survey of the animal kingdom discloses at least 40 orders containing one or more forms known to produce light and several more orders containing species whose luminosity is doubtful. In the plant kingdom there are two groups containing luminous forms, the bacteria and the true fungi. Among animals the best known are the flagellates; hydroids; jellyfish; ctenophores; sea pens; *Chaetopterus* and other marine worms; earthworms; brittle stars; various crustaceans; myriapods; fireflies and glowworms; *Pholas dactylus* and *Phyllirrhoë bucephala*, both molluscs; squid; *Pyrosoma*, a colonial ascidian; and many fishes. These will be considered in more detail in Chapter II.

Apparently there is no rhyme or reason in the distribution of luminescence throughout the plant or animal kingdom. It is as if the various groups had been written on a blackboard and a handful of damp sand cast over the names. Where each grain of sand strikes, a luminous species appears. The coelenterates have received most sand. Luminescence is more widespread in this phylum and more characteristic of the group as a whole than any other. Among the arthropods, luminous forms crop up here and there in widely unrelated groups. In the molluscs, excluding the cephalopods, only three luminous species are known. Several large groups contain no luminous forms whatever. It is an extraordinary fact that one species in a genus may be luminous and another closely allied species contain no trace of luminosity. There seems to have been no development of luminosity along direct evolutionary lines, although a more or less definite series of gradations with increasing structural complexity may be traced among the forms with highly developed luminous organs.

Luminous animals are practically all either marine or terrestrial. No examples of fresh-water luminous organisms are known except bacteria and an aquatic glowworm. Of marine forms, the great majority are deep-sea animals, and it is among these that the development of true luminous organs of a complicated nature is most pronounced. Many of the luminous marine animals are to be found floating at the surface, the plankton, while the littoral luminous forms are in the minority.

COLD LIGHT

Not only the adults but the embryos and even the eggs of some animals are luminous. It was known to the older investigators that freshly laid eggs of the fireflies are quite luminous. The light does not come from luminous material of the luminous organ adhering to the egg when it is laid, but from within the ovarian egg itself. The early segmentation stages of ctenophores are luminous on stimulation, but the unsegmented eggs do not luminesce. Larvae of some crustaceans, worms and brittle stars also produce light.

It is not surprising to find many false reports of luminous animals in the literature and we cannot be too careful in accepting as luminous a reported case. The difficulty lies chiefly in the fact that all luminous organisms with the exception of bacteria, fungi and a few fish, flash only on stimulation, and, while it is easy enough to see the flash, the animal is lost between the flashes. Anyone who has ever tried to determine what animal is responsible for the occasional flashes of light observed on agitating almost any sample of sea water will realize how difficult it is to discover the luminous form among a host of non-luminous ones, especially if the animal is microscopic in size. There is in addition the possibility of confusing the light from a supposed luminous form with the light from truly luminous organisms living upon it. The reported cases of luminosity among marine algae are now known to be due to hydroids or unicellular organisms living on the alga.

Many errors have resulted from attempts to infer luminescence in an animal by study of preserved material. The most celebrated case is the candle-fly or lantern-fly (*Fulgora*), a beetle of tropical countries whose name suggests light and whose head and thorax is like a lantern in shape. This beetle was reported as luminous by Grew in 1681 from study of a dried specimen. Since that time numerous statements of luminescence have appeared and it has frequently been confused with the brilliantly luminous beetle, *Pyrophorus*. The best evidence indicates that *Fulgora* is not self-luminous.

We know also that many non-luminous animals may become infected with luminous bacteria, not only after death but also while living, so that their luminescence is purely secondary. Giard and Billet (1889, 1890) discovered living luminous sand fleas whose body had become completely infected with luminous bacteria, and succeeded in inoculating many different kinds of amphipod and isopod crustaceans with them, in some cases passing the infection from one to the next through nine individuals. The sand fleas might easily have been taken for truly luminous animals if not carefully investigated. Luminous bacterial maladies of living mayflies, midges, mole crickets, caterpillars, and fresh water shrimp have also been described and in many cases the special bacteria isolated and luminous colonies produced on culture media. For these forms the infection is fatal.

By inoculating sea animals with ordinary luminous bacteria, it is often possible to make them luminous. The animal may finally die or it may survive. Tarchanoff (1901) injected luminous bacteria into the dorsal lymph sac of frogs with the result that the animals continued to glow for three or four days, especially about the tongue. The author has observed luminous frogs in Cuba but it turned out that these animals had just finished a hearty meal of fireflies, whose light was shining through the belly with considerable intensity.

The most interesting forms are those in which luminous bacteria are always present in every individual throughout its life. Frequently a special organ has been developed to harbor the bacteria and they can with certainty be considered symbiotic. Truly remarkable cases are the luminous fishes of the Banda Sea, *Photoblepharon* (Fig. 7) and *Anomalops*, whose light organ, just below the eye and specially designed for growing masses of luminous bacteria, has a rich blood supply, opaque screens to protect other tissues of the fish from the light and a mechanism for turning the light on and off. These fish, together with a number of others, will be described in Chapter II. Only a careful microscopic observation of the organ would reveal the bacterial origin of its light.

Indeed, at one time Pierantoni (1914), Buchner (1914) and Zirpolo (1918) supposed that the light of practically all luminous animals is of symbiotic bacterial origin, and in such forms as the fireflies, whose luminous cells do contain granules, that these granules are really bacteria which have become so modified that they are no longer typical in appearance and cannot be grown on artificial culture media. They are believed to be transmitted from one generation to the next in the egg and to multiply in the cells of the firefly larva. Although there are undoubtedly symbiotic bacteria living in some luminous animals, the behavior of the photogenic granules in the great majority of cases is so different from the behavior of unquestioned luminous bacteria that it seems an exaggeration to consider that *all* bioluminescence is bacterial in origin.

Apart from these animals which actually luminesce but whose light is bacterial and secondary, not produced by the animal itself, there are many forms whose surface is so constituted as to give interference colors in a dim light. Some birds and butterflies, whose feathers and scales are iridescent, have been erroneously described as luminous, as have also the eyelike spots of the long spined sea urchin, *Diadema setosum*, as well as similar structures of the starfish, *Brisingia coronata*. Perhaps the best known case among aquatic animals is *Sapphirina*, a marine copepod living at the surface of the sea, and especially likely to be collected with other luminous forms. Its cuticle is so ruled with fine lines as to diffract light and to flash on moving, much as a fire opal. Needless to say no trace of light is given off from this animal in a totally dark room. Luminescence must not be confused with iridescence.

It has often been supposed that the eye of a cat (or of other animals) is luminous. Pliny and Francis Bacon (1620) mention this, among other luminescences. The eyes of a moth, also, can be seen to glow like beads of fire when it is flying about a flame. All of these cases are, however, purely reflection phenomena and due to reflection, out of the eye again, of light which has entered from some external source such as the headlight of an automobile. The eye of any animal is quite invisible in absolute darkness. However, the lens of the eye is markedly fluorescent in ultraviolet light and there always appears a peculiar light haze from the fluorescent lens when one observes a mercury lamp through a filter transparent to ultraviolet but opaque to visible radiation.

Some green plants have been reported as luminescent but this appearance is again due to reflection. The moss, *Schistostega osmundacea*, which lives in caves and dimly illuminated places, has cells which are almost spherical, constructed like a lens, so as to refract the light and condense it on the chloroplasts at the bottom of the cells. Some of the dim light is reflected out of the cells again and gives the appearance of self-luminosity. The alga, *Chromophyton rosanoffi*, is another example of apparent luminosity, due to reflection from almost spherical cells.

We are probably to consider as reflectors also the row of spots along the sides of a male lizard, *Proctoporus Shrevei*, from Trinidad, recently described as luminous organs by H. W. Parker (1939). The evidence is not yet sufficient to class these pearly spots as self-luminous, although the animal lives in dim, secluded places. It is interesting to find that many years ago a reptile could develop a series of reflector buttons quite similar to modern markers along the roadside, which reflect the light of a passing car.

It is said (Mornay, 1816) that the white latex or milky juice of *Euphorbia phosphorea* from Brazil can be used to write on paper, giving characters which appear luminous in the dark. Luminescence of a white surface is always questionable, since the reflection of such a dim light as starlight may give the appearance of light emitted. It is easy to understand that the white feathers of some herons, the white tufts on nestling birds which live in caves, and the white beak of the Australian finch (see Chun, 1904) may appear luminous by reflection, but the reports of luminous barn owls in England are decidedly mysterious. Perhaps they have been mistaken for white Arctic owls or have become infected with luminous fungus from wood about their holes.

It has been suggested that the phenomenon of the "purple arcs" is due to luminescence. These are two curved purplish or bluish bands of light easily seen when a vertical slit of red light is observed in a dark room. The projection of these arcs back on to the retina shows that they follow the course of nerve fibers from the fovea, the point of central vision, to the blind spot, the entrance of the optic nerve. Ladd-Franklin (1926) believed the purple arcs were the result of luminescence of nerve fibers carrying impulses, stimulated by the red light, to the brain. The luminescent nerve fibers would be so close to the underlying rods of the retina that these rods would be affected and the sensation would be that of the purple arcs. However, no luminescence of nerve fibers carrying nerve impulses has ever been detected, despite careful observation. The true explanation is probably stimulation of underlying retinal rods by the electric action potentials of nerves responding to the red light. Stimulation of a retinal rod by any means would give the sensation of light.

LIVING LIGHT

The flashing of flowers, yellow or orange in color, has been described by many persons, including the daughter of Linnaeus. A vivid account has been given by Phipson (1870, pp. 79-86). There is no light in complete darkness but at sunset the flowers seem to "emit sparks" or "throw out flames." These have been regarded either as electrical discharges or a subjective phenomenon due to formation of after images in partially dark adapted eyes. Rupture of the spathe which encloses the flowers of *Pandanus* is accompanied by a cracking noise and a spark of light. Such a luminescence might easily be an electrical discharge.

There are several such light phenomena known which have nothing to do with living organisms. Commonest of these is St. Elmo's fire ("corposants" of English sailors), a glow accompanying a brush discharge of electricity, which appears as a tip of light on masts of ships, spires of churches, trees or even the fingers of the hand. It is best seen in winter, during and after snowstorms and is a purely electrical phenomenon, although a case of luminescence. Electrical discharges may occur from the hats and clothing of mountaineers when conditions are just right. Stroking the fur of a cat or currying a horse often results in a brush discharge.

Many observers have reported luminescent rain drops. Some of these reports have been collected by Francis Arago, the French physicist and astronomer, and are quoted from Phipson (1870, p. 45):

"In 1761, Bergman, the celebrated Swedish chemist, wrote to the Royal Society of London that he had observed on two occasions, towards evening, and when no thunder was heard, rain which sparkled as it touched the ground, *making the latter appear as if covered with waves of fire.*

"On the 3rd of May, 1768, near Arnay-le-Duc, M. Pasumot *was overtaken on an open plain by a violent storm. The rain-*

water collected abundantly on the border of his hat; and when he stooped his head to let it flow off, he observed that, in its fall, encountering that which fell from the clouds, at about twenty inches from the ground, sparks were emitted between the two portions of liquid.

"On the 28th of October, 1772, on his way from Brignai to Lyons, the Abbé Bertholon was caught in a storm at five o'clock in the morning. Rain and hail fell heavily. The drops of rain and the hailstones which struck the metallic parts of the mounting of his horse's trappings, emitted jets of light."

Some of these descriptions sound like reflected light from spherical particles but electric discharges do occur from hail or rain drops. Falling snowflakes were observed to luminesce by Ghave (1856).

Waterspouts at night or dust from volcanic eruptions, may sometimes be luminous. Luminous clouds and fogs have frequently been described (see Phipson, 1870, pp. 54-61) but it is impossible to say whether a luminescence has been observed, or reflection from some other light. The early descriptions are quaint but not properly controlled. The aurora borealis (and australis) is a true luminescence on a grand scale.

Less well known is the *Ignis fatuus*, also called Jack-o'-Lantern, Spunkie, Will-o'-the-Wisp, elf candles, or corpse candles of Welch graveyards. It appears in low grounds, over marshes and stagnant pools, as a pale bluish flame which may be fixed or in motion, steady or intermittent. So uncommon is this phenomenon that its nature is not well understood, but it has been attributed to burning phosphine, a self-inflammable gas, generated in some way from the decomposition of organic matter in the swamp. Unfortunately, phosphine is not known as a decomposition product of organized matter. Methane, a well known decomposition product and abundantly formed in swamps, will burn with a pale bluish flame and some have thought the *Ignis fatuus* to be the result of this gas. As methane is not self-inflammable there remains the difficulty of explaining how it becomes ignited. Although still a mystery, it is possible that this light is also of electrical origin or that in some cases large clusters of luminous fungi or swarms of luminous gnats have been observed. Perhaps a wisp of fog hanging over a swamp and seen in moonlight or starlight is the simplest explanation of this "foolish fire."

What are we to say of a Will-o'-the-Wisp described by Dr. Priestley—and quoted from Phipson (1870, p. 68). "A gentleman who had been making electrical experiments for a whole afternoon in a small room, on going out of it, *observed a flame following him at some little distance*." Here there seems to have been a difference between the artificial *Ignis fatuus* and that met with in nature, for the flame *followed* the gentleman as he went out of the room, but the natural phenomenon generally *recedes* as we approach it.

There are some cases of luminosity on record in connection with man himself. Before the days of aseptic and antiseptic surgery, wounds frequently became infected with luminous bacteria and glowed at night. The older surgeons even supposed that luminous wounds were more apt to heal properly than non-luminous ones. We know that luminous bacteria are non-pathogenic, harmless organisms and the presence of these forms on dead fish or flesh never accompanies but always precedes putrefaction. No harm has come from eating luminous meat, unless it may also have become infected with pathogenic forms.

A few cases of luminous individuals have been noted in which the skin was the source of light, especially if the person sweated freely. It is possible that here we are again dealing with luminous bacteria upon the accumulations of substances passed out in the sweat, which serves as a nutrient medium.

COLD LIGHT

Priestley in his *History of Optics* (p. 587) records a case where the sweat "*has been observed to be phosphoraceous, without any preparation. This once happened to a person who used to eat great quantities of salt, and who was a little subject to the gout, after sweating with violent exercise. Stripping himself in the dark, his shirt seemed to be all on fire, which surprised him very much.*"

There are also on record, in the older literature, cases of luminous urine, where the urine when freshly voided was luminous (Guyton-Morveau, 1815; Driessen, 1818) or a jar of urine became luminous on stirring. The urine of the civet cat is also said to light (Azara, 1801; Langsdorff, 1812). If these observations are correct, we are at present uncertain of the cause of light. Bacterial infections of the bladder are not inconceivable although most luminous bacteria are strongly aerobic and would not thrive under anaerobic conditions. Luminous bacteria will live in normal human urine, but not well. In diabetic or albuminous urines, it is very likely that they might thrive better, and it is possible that the luminous urines reported are the result of luminous bacterial infection. On the other hand, the light may be purely chemical, due to the oxidation of some compound, an abnormal incompletely oxidized product of metabolism, which oxidizes spontaneously in the air. We know that sometimes these errors in metabolism occur, as in *alkaptonuria*, where homogentistic acid is excreted in the urine and on contact with the air quickly oxidizes to a dark brown substance. Light, however, has never been reported to accompany the oxidation of homogentistic acid, although it does accompany the oxidation of some other related organic compounds. (See Chapter III.)

When we sift the true from the inferred records of luminescence, we discover a wealth of interesting material, especially in the living world. In the following chapters we shall

LIVING LIGHT

deal with the structure of luminous organs, methods of lighting, the purpose of the light, types of luminescence, chemistry of the process, the intensity, spectra and efficiency of luminescence, and the possibility of imitating the light of living things.

CHAPTER II

LIGHT-PRODUCING ORGANISMS

LUMINOUS organs have often been mistaken for eyes. The reason is obvious; they are eyes in reverse. Chemical *production* of light is the converse of the chemical *detection* of light. The lantern of fishes is an organ of chemiphotic change; the eye an organ of photochemical change. Moreover, photochemical reactions and chemiluminescent reactions have this in common, that they are largely but not exclusively oxidations. In light production by living things some material is oxidized.

In general, we may divide luminous organisms into two great classes according as to whether the oxidizable material is burned within the cell where it is formed or is secreted to the exterior and is burned outside—intracellular and extracellular luminescence. Animals with intracellular luminescence have the most complicated luminous organs.

Even the evolution of complex light organs and eyes is similar. In the simplest unicellular forms, certain pigment granules within the cell serve as the photochemical detectors of light, while in luminous protozoa, similarly, photogenic granules scattered throughout the cell are oxidized with light production. In the higher forms, the eye contains groups of photosensitive rods and cones connected with afferent nerves, lenses, and accessory structures for properly adjusting the light, while luminous organs contain groups of photogenic cells in connection with efferent nerves, lenses, and accessory structures for properly directing the light. In the two groups where the eye has attained its highest development, the cephalopods and vertebrates, the luminous organ is also found in greatest com-

LIVING LIGHT

plexity and perfection. In intermediate stages of evolution the eye and luminous organ so closely approach each other in structure that it was a much debated question whether certain organs found in worms and crustacea were intended for receiving or producing light.

Luminescence appears usually to be associated with granules. In organisms with extracellular luminescence they are contained in various forms of glands, either scattered single gland cells or multicellular types of all degrees of complexity. Light appears when these granules are secreted. Sometimes more than one type of gland cell can be distinguished, corresponding to the existence of two substances necessary for light production—luciferin, a relatively simple heat stable, easily oxidizable compound, and luciferase, an enzyme which accelerates its oxidation. These two substances, first demonstrated by Dubois in 1885, have been discovered in a number of luminous animals but by no means in all. The mechanism by which their interaction produces light, their properties, and their distribution among light-producing organisms will be discussed in Chapter IV, devoted to the chemistry of light production.

We may also divide luminous forms into two groups according as to whether the oxidation of luminous material goes on continuously, independently of any stimulation of the organism; or is intermittent, oxidation and luminescence occurring only as a result of stimulation, using the word "stimulation" in the same sense in which it is used in connection with nerve or muscle tissue. Bacteria, fungi, a few fish (containing symbiotic luminous bacteria) and the beetle *Phengodes* produce light continuously and independently of stimulation. Its intensity varies only over long periods of time and is dependent on the nature of the nutrient medium or general physiological condition of the organism. All other forms give off no light until they are stimulated. Stimulation may of course come from within (nerves) or from without the animal. Only under favorable conditions, such as will eventually lead to the destruction of the luminous cells, do these forms give off a continuous light. This has often been spoken of as the "death glow," and is to be compared with *rigor* in muscle tissue. The means by which stimulation sets off the luminescence will be discussed in Chapter V on the physiology of light production.

In the following pages the structure of luminous organs will be considered in the bacteria, fungi, radiolarians, flagellates, sponges, hydroids, jellyfish, siphonophores, sea pens, ctenophores, corallines, marine worms, earthworms, brittle stars, crustaceans, myriapods, insects, spiders, molluscs, squid, balanoglossids, ascidians and fish. This strikingly diverse array of luminous creatures is distributed in most of the phyla or great subdivisions of the animal kingdom. Three phyla, the platyhelminthes, the nemahelminthes and the trochelminthes, as the names indicate, the flat worms, thread worms and wheel worms, are frequently parasitic. They contain no known luminous animals. The plant kingdom has only two luminous groups, bacteria and fungi.

LUMINOUS BACTERIA

The smallest lamps in the world, luminous bacteria, are no different from ordinary bacteria except in their ability to luminesce. They are so small that the light from a single one cannot be seen by the eye, with or without a microscope. It takes many thousands together to produce enough light to affect the retina. An individual of typical form is a short cylinder, a bacillus measuring $1.1 \times 2.2\mu$,[1] the volume about 1/1,700,000,000 of a cubic millimeter (Fig. 8). Others are nearly spherical, cocci, while still others are curved rods,

[1] A micron $(\mu) = 0.001$ millimeter or 1/25,000 inch.

vibrios, related to the cholera organism. They are mostly motile, possessing one or more flagellae. Many different species of phosphorescent bacteria have been described, differing in cultural characteristics and structural peculiarities, and grouped in the genera, *Bacterium, Photobacterium, Bacillus, Coccobacillus, Achromobacter, Micrococcus, Microspira, Pseudomonas,* and *Vibrio*. Specific names indicating their light-producing power, such as *phosphorescens, phosphoreum, luminosum, lucifera,* etc., have been applied.

Luminous bacteria ordinarily live on meat but few persons before the early part of the nineteenth century suspected the cause of the light. One of the earliest observers was the anatomist, Fabricius ab Aquapendente at Padua, who noted luminous mutton in 1592, as did also the Danish philosopher, Bartolin in 1641. Boyle (1672) studied luminous calf, pig and chicken flesh, while Beal (1676) observed fish and Paullinus (1707) saw luminous hen's eggs. Phosphorescent light may appear on the dead body of a man. Such cases are said to be not infrequent in dissecting rooms but escape observation since students or professors are unlikely to visit them at night. Meat hanging in cellars or refrigerators is apt to become infected, and early experiments, mostly on the effect of gases and other substances on luminescence, are associated with the names of Martin (1764), Canton (1769), Hulme (1800, 1801), Hermbstadt (1808), Dessaignes (1809), Heller (1853), Hankel (1862), Pflüger (1875), Nuesh (1877), Lassar (1880), Ludwig (1884) and Fischer (1888).

Some strains live in ordinary pond or tap water but many are marine and require a high salt content. They have originally come from dead fish or from some sea creature. They do not, however, contribute to the phosphorescence of the sea since their number in ordinary sea water is too small. It is only when they grow in great numbers in some nutrient material that we observe their light. Then the whole medium shines in the dark, an effect which so greatly intrigued Robert Boyle and the early observers. Despite the close resemblance of some luminous bacteria to the cholera *Vibrio*, they are quite harmless. They can easily be grown at room temperature on 1.5 per cent agar jelly in sea water containing 2 per cent peptone, 1 per cent glycerine and some powdered calcium carbonate, added to maintain the proper alkalinity of the culture medium (Fig. 8). When scraped off the agar-jelly and suspended in sea water, the dense emulsion forms a cold greenish liquid fire. It must be continually shaken to dissolve more oxygen, otherwise the light will go out, remaining visible only at the surface where the bacteria are in contact with air.

Beijerinck (1889, 1902) was the first to utilize bacteria as a test for oxygen. By mixing luminous bacteria with an emulsion of chloroplasts (from clover leaves) in the dark, allowing the bacteria to use up all the oxygen, and then exposing the mixture to light of various colors, the effect of different wave lengths of light on photosynthesis could be studied. Only if the chloroplasts are exposed to a color in the spectrum which decomposes carbon dioxide with liberation of oxygen do the bacteria luminesce. When this oxygen is used up by the bacteria, the tube becomes dark. Merely striking a match is sufficient to again form oxygen.

Molish (1904) also used luminous bacteria to show that dried and powdered leaves placed in a suspension of the bacteria could produce oxygen when illuminated and the author (1928) has studied the evolution of oxygen by various green, red and brown algae kept in absence of oxygen for from 30 minutes to several hours. A species of *Fucus*, when illuminated after absence of oxygen for four hours, immediately produced an abundance of this gas which allowed the bacteria to luminesce. The first step in photosynthesis can therefore occur in complete absence of oxygen. Hill (1928) has used luminous bacteria to study the penetration of oxygen through rubber, collodion, vaseline and other substances.

Beijerinck (1889, 1902) tested bacterial filters by means of luminous bacteria. If there is a crack in the filter the bacteria will pass through and a luminous filtrate is the result, but a perfect filter allows no organisms to pass and gives a dark filtrate.

It is almost impossible to make out structural differences within a bacterial cell and we cannot definitely state in just what special region, if any, the luminescence is produced. Kishitani (1928, 1933) has made recent studies. No photogenic granules have ever been certainly observed. As Beijerinck showed, filtration of the bacteria from their culture medium gives a dark sterile filtrate absolutely free from any luminous secretion. This is also true if the filtration is carried out in absence of oxygen to prevent oxidation of luminous substance. Readmission of oxygen to the filtrate gives no light but readmission to the bacteria gives a brilliant luminescence. Luminescence is undoubtedly intracellular in bacteria.

Luminous bacteria can be cooled to the temperature of liquid air ($-190°C$) and will luminesce and live when warmed again (Macfayden, 1900). But if thoroughly ground up at that temperature, there is no luminescence when again warmed (Macfayden, 1902). These bacteria are not spore formers and do not live in the dry state for any length of time. If dried quickly over calcium chloride they can luminesce and some will grow when again moistened. However, if ground up in the dry state and then moistened, no light appears. Luciferin and luciferase have never been certainly demonstrated.

Luminous bacteria are best kept permanently in sealed tubes in a hydrogen or nitrogen atmosphere (Shoup, 1928) where they may exist indefinitely. They will grow whenever oxygen is admitted. In this way a new colony can always be started and will soon increase to the point where its light is sufficient to read by or to photograph. A few species are facultative anaerobes and can grow in absence of oxygen. Almost every worker with luminous bacteria has devised a "living lamp," a large flask of luminous bacteria, and taken such photographs as are illustrated in Fig. 9.

Biologically, luminous bacteria are of most interest through their relation to living animals. Reference has already been made in Chapter I (p. 14) to Giard and Billet's (1889, 1890) discovery of luminous sand fleas (Fig. 10) which have become infected with luminous bacteria scattered throughout the body, especially in the muscles, giving rise to a malady which eventually leads to their death. Although the bacterium sometimes produced no light on artificial culture media, light always appeared when inoculated into other crustaceans. The author has seen and Inman (1927) has described a similar occurrence at Woods Hole, Mass., and has grown the organism, *Bacterium Giardi*, on culture media. A bacteriological examination of twenty non-luminous sand fleas showed that luminous bacteria could be isolated from the intestinal tract in almost every case. It is quite possible that at certain times or under certain conditions these bacteria invade the muscles and spread so rapidly that the sand fleas become luminous and die.

A large number of other animals have been described, both fresh-water and land forms, in which luminous bacterial infections result in luminescence while the animal is living. This has been observed in caterpillars (Gimmerthal, 1829; Boisduval, 1883; Stammer, 1930; Pfeiffer and Stammer, 1930). The bacterium (*B. haemophosphoreum*) is different from other luminous bacteria and will cause a luminous infection in other insects but not in higher forms. Mayflies (Hazen, 1873; Eaton, 1880, 1882) and mole crickets (Ludwig, 1891) also become luminous at times, presumably through luminous bacteria. Possibly the luminous termite hills observed by Knab (1909) in Brazil contained termites infected with bacteria and the same may be said for occasional records of luminous spiders (Brown, 1925) or for the luminous ant, described by Wheeler (1916).

The fresh-water shrimp of Lake Suwa in Japan frequently become luminous from a bacillus called *Microspira phosphoreum* (Yasaki, 1927). This form can be grown on nutrient agar containing 0.5 per cent sodium chloride and produces brilliantly luminous colonies. They are toxic for fish, birds, mice and guinea pigs but not for rabbits or man. Immune reactions of other bacteria have been described by Hinomiya (1924) and Meissner (1926).

In all the above instances, the bacterial infection eventually leads to the death of the animal. The bacteria are spoken of as parasitic, in contrast to the saprophytic forms which grow on dead fish or meat. Then there are also symbiotic bacteria, cases in which animals are permanently luminous as a result of harmless luminous bacteria always present and living in a definite structure, usually of glandular form and frequently specially designed to culture the bacteria. This is true symbiosis.

Osorio (1912) first described bacteria in a large open gland in the fish, *Malacocephalus*, but the observation was so unique that little attention was paid to the discovery. The Portuguese fishermen smear the luminous secretion containing bacteria on their bait to make it attract other fish.

Pierantoni (1914) and Buchner (1914), a specialist on intracellular organisms, emphasized the possibility of bacterial symbiosis in the fireflies[2] and *Pyrosoma* (Fig. 14) and were at first inclined to believe that all luminescence in animals is due to symbiotic bacteria. Since practically all luminous cells contain granules or rod-shaped bodies it might be that these are bacteria so changed by symbiotic existence that their bacterial structure is largely lost. Fireflies are a case in point but firefly-bacterial symbiosis is very doubtful. (See p. 72.) According to Buchner (1930) who has written an excellent monograph on the subject, there can be no doubt of bacterial symbiosis in *Pyrosoma* and probably in *Salpa*. In earthworms, medusae and ctenophores, it is also very doubtful if the light is of bacterial origin, but squid and fish, particularly, contain undoubted symbiotic forms. Zirpolo (1917-1938) has studied the physiological characteristics of many of these bacteria.

The most striking examples of luminous symbiosis are the two East Indian fish, *Photoblepharon palpebratus* and *Anomalops katoptron*, 8 to 11 cm. long, remarkable in many respects (Fig. 12). *Photoblepharon* is found only in the Banda Islands in the very middle of the archipelago. It was first described by Boddaert (1781). *Anomalops* is abundant at Banda, but has occasionally been taken in Celebes, Fiji, New Hebrides, the Paumotus and Puerto Rico. It was described by Bleeker in 1858. Both fish have been studied by Steche (1909) but the bacteria were first described by the author (1921). They can be easily seen with a microscope. As the names indicate (*Photoblepharon* = light eyelid, and *Anomalops* = irregular eye), and the figures show, there is a large white organ just under the eye of each fish. It is not always visible, for the fish are able to conceal it at will, and different methods of concealment are used in the two cases. *Photoblepharon* has a fold of

---

[2] Dubois (1887) spent much time trying to grow bacteria from light organs of luminous animals and Kuhnt (1907) thought lampyrids contained luminous bacteria.

LIVING LIGHT

black tissue on the lower surface, which can be drawn up over the organ like an eyelid, completely obscuring it. In *Anomalops* the organ is attached at the antero-dorsal corner by a hinge which allows the whole organ to be turned over and downward into a groove or pocket, so that none of the white surface is visible.

We now know that the white structure is a luminous organ and produces light, but the earlier views of its function, arrived at merely by inspection of preserved material, were indeed fantastic. Some thought that the black screen in *Photoblepharon* was to protect the eyes of the fish from injury by the branches of the coral among which the fish lived, while others thought it a protection of some delicate tissue against the rays of the tropical sun. Any native could have revealed the function of this tissue, for the Banda islanders cut out the luminous organ, attach it to a hook, and use it for bait in fishing. The light lasts for seven or eight hours.

The Banda Islands rise from the Banda Sea, a very small volcanic group, whose peak, Gunong Api, has been active during the past century. The harbor of Banda Neira is almost completely landlocked and formed an excellent anchorage for Dutch vessels in the rich days of the spice trade. For Banda was once a prosperous place where wealthy Dutch merchants exported nutmegs to all the world. It still is a nutmeg center and old villas still remain, but they are empty and the air is one of isolation, of peace and quiet and decay.

The coastal waters are rich with corals but the bottom drops suddenly to depths of 12,000 feet. However, these fish are not deep-sea forms. *Photoblepharon* swims alone or a few will swim together among the stones and corals. Hence the native name, *ikan* (fish) *leweri* (light?) *batu* (stone). *Anomalops* swims in schools of a hundred or more at the surface but in somewhat deeper water. Hence the native name, *ikan leweri*

LIGHT-PRODUCING ORGANISMS

laut (sea). It is a most extraordinary sight to watch these fish swimming through the water, turning their lights now on and now off, like great marine fireflies. The actual light production is continuous since it is due to luminous bacteria.

The organ in each fish is made up of rows of tubes (Fig. 13) with blood capillaries running between them and a very rich blood supply, for the organ is extremely sensitive to lack of oxygen and becomes dark very quickly when the supply of oxygen is discontinued. If the contents of the luminous tubes are examined with a microscope, a mass of moving bacteria, curved rods in most cases, can be seen.

Although the author made many attempts to grow the bacteria from *Photoblepharon* in artificial culture media, no success was attained. Bacteria from the organ grew in artificial media but they never produced light. This may be due to the fact that these symbiotic forms require some peculiar nutrient material of the living fish which is absent in the artificial media. However, symbiotic bacteria from other forms have been cultured.

Whenever the luminous organ of an animal produces a steady light we may suspect the presence of luminous symbionts. Several other widely different kinds of fish belong in this category. They possess an open gland which harbors luminous bacteria. These can be easily grown on artificial culture media. In the Japanese king-fish, *Monocentris japonica* (Fig. 15), which has a gland on the lower jaw,[3] they have been isolated as slightly curved rods (Yasaki, 1928); in the fish *Acropoma japonicum* they have the form of cocci and are called *Coccobacillus acropoma* (Yasaki and Haneda, 1936). Here the organ is embedded in muscles of the ventral wall and has a long duct. In *Physiculus japonicus* (Kishitani, 1930) the organ is also ventral (Fig. 16) and similar to that of *Coe-*

---

[3] Okada (1926) overlooked the bacteria in *Monocentris*.

lorhynchus (Hickling, 1931) and *Malacocephalus* (Hickling, 1925, 1926).

In this fish Hickling reported the luciferin-luciferase reaction and believed the granules were not bacteria, although they show some of the properties of living cells. Haneda (1938), however, proved definitely that the light was of bacterial origin by growing the bacteria in culture media. *Coelorhynchus japonicus* and *Hymenocephalus atriatissimus*, of the same family, also contain luminous bacteria. Although the gland in these fish is well hidden under the skin some light shines through and this can be increased and decreased by contraction and expansion of chromatophores. The isolated bacteria of *Physiculus japonicus* are called *Micrococcus physiculus*. Dahlgren (1928) has described an open gland on the tip of the anterior dorsal fin ray of *Ceratias* which appears from histological sections to contain bacteria, and Harms (1928) found luminous bacteria in the open gland of the East Indian fish, *Equula*.

In all these cases, every fish always contains an organ which is luminous and the bacteria are often culturally quite different from ordinary luminous bacteria. The organ is frequently specialized and may contain structures that could only be of value in connection with light production, for example, the screen of *Photoblepharon*. A true symbiosis is indicated. As a result of this partnership the bacteria obtain a perfect nutrient medium, the fish a lantern, a warning signal or a recognition mark, whatever its use may be.

It is much more difficult to interpret the luminous bacteria found in squid. Pierantoni (1918) and Zirpolo (1918) first described the luminescence of the squid, *Sepiola intermedia* and *Rondeletia minor* as a true hereditary symbiosis. In such squid there can be no doubt of the presence of luminous bacteria in the accessory nidamental glands. The light may shine through the mantle or may be so faint that it is only visible when the mantle is dissected off. According to Meissner (1926), these bacteria are different from ordinary bacteria and show marked strain specificity and special relations to their hosts. Kishitani (1928) has described open luminous organs containing bacteria in the myopsid squid, *Loligo edulis*, *Sepiola birostrata* and *Euprymna morsei*. In *Loligo* the organ is paired and consists of a lens, reflector and secretory tubules filled with a new luminous bacterium, *Coccobacillus loligo*. The bacteria are not transferred from generation to generation through the host but come in from the outside. In the remarkable deep-sea squid *Heteroteuthis dispar*, luminous bacteria have been suspected (Pierantoni, 1924) but the situation is far from clear (Mortara, 1924; Skowron, 1926). The European forms have been reinvestigated by Herfurth (1936).

The problem is one of interpretation. Skowron (1926) examined thirty individuals of *Sepiola intermedia* and found thirteen that showed no trace of luminescence. He concluded that the luminous bacteria often found in these squid were not true symbionts or they should be always present. The readiness with which various marine forms can be infected has also been demonstrated by Skowron. *Paleamon*, *Amphioxus*, *Sepia* and many fish can be made luminous by injecting cultures of ordinary marine forms. In *Ciona* they grow beautifully in the ovary. The injected forms may die or they may recover completely and the luminescence disappear slowly over a period of two or three weeks. As long ago as 1887, Dubois found luminous bacteria in the siphon of *Pholas* but decided they were only occasional guests (1890).

The rod and spindle shaped bodies of the light organ of the Japanese hotaru-ika (firefly squid), *Watasenia scintillans*, were thought to be bacteria by Shima (1927), who claimed to have grown colonies of luminous bacteria from them. However, these findings have been denied by Hayashi (1927), Kishitani (1928), Okada, Tagaki and Sugino (1933) and Tagaki (1933). Kishitani (1932) is of the opinion that all the oegopsid squid, such as *Watesenia, Abralia, Enoproteuthis* and *Chiroteuthis*, whose closed organs contain granules or rods like bacteria, are really self-luminous and the granules are photogenic granules.

It is perhaps best not to generalize too widely regarding luminous symbiosis but to recognize that some animals may occasionally become infected with luminous bacteria and die; some possess glands so advantageously placed that they easily become infected locally without the death of the animal; in other forms certain glands always contain local luminous bacteria. Finally in the course of evolution accessory structures develop in connection with the gland and true symbiosis appears.

If bacteria have developed the ability to produce light, it is logical to believe that cells of protozoa and multicellular animals could also. There is no reason *a priori* to attribute all luminescence to modified bacteria; in fact the chemical behavior of the photogenic granules of many animals indicates they could not possibly be modified bacteria. It is these latter which are responsible for the brightest luminescences in the animal kingdom.

LUMINOUS FUNGI

No summer passes without the discovery of some phosphorescent wood, the fox-fire of English legend, often associated with elves and fairies. Old stumps of trees, moulding leaves, the inside of bark from logs on the ground will glow brilliantly and continuously, although actually cold. Most observers report keeping glowing chips of wood for days or weeks. On October 29, 1667, Robert Boyle, carrying out experiments with his air pump, wrote:

"Exp. I: *Having procured a Piece of shining Wood, about the bigness of a groat or less, that gave a vivid Light, (for rotten Wood) we put it into a middle sized Receiver, so as it was kept from touching the Cement; and the Pump being set a-work, we observed not, during the 5 or 6 first Exsuctions of the Air, that the splendor of the included Wood was manifestly lessened (though it was never at all increased;) but about the 7th Suck, it seemed to glow a little more dim, and afterwards answered our Expectation, by losing of its Light more and more, as the Air was still farther pumped out; till at length about the 10th Exsuction, (though by the removal of the Candles out of the Room, and by black Cloaths and Hats we made the place as dark as we could, yet) we could not perceive any light at all to proceed from the Wood.*

"Exp. II: *Wherefore we let in the outward Air by Degrees and had the pleasure to see the seemingly extinguished Light revive so fast and perfectly, that it looked to us almost like a little Flash of Lightning, and the Splendor of the Wood seemed rather greater than at all less, than before it was put into the Receiver.*"

The experiment shows very clearly the necessity of air for luminescence. If a glowing stump is chopped with an axe the chips are luminous throughout and look like so many glowing coals. Boyle was impressed with the similarity of the light giving process in glowing coal and shining wood as he draws a comparison between the two which brings out the fundamental similarity of combustion processes.

"Resemblances:

VII. *The Things wherein I observed a Piece of shining Wood and a burning Coal to agree or resemble each other are principally these five:*

LIVING LIGHT

"1. Both of them are Luminaries, that is, give Light, as having it (if I may so speak) residing in them; and not like Looking-glasses, or white Bodies, which are conspicuous only by the incident Beams of the Sun, or some other luminous Body, which they reflect. . . .

"2. Both shining Wood and a burning Coal need the Presence of the Air (and that too of such a Density to make them continue shining. . . .

"3. Both shining Wood and a burning Coal, having been deprived, for a Time, of their Light, by the withdrawing of the contiguous Air, may presently recover it by letting in fresh Air upon them. . . .

"Both a quick Coal and shining Wood will be easily quenched by Water and many other Liquors. . . .

"5. As a quick Coal is not to be extinguished by the Coldness of the Air, when it is greater than ordinary; so neither is a Piece of shining Wood to be deprived of its Light by the same Quality of the Air. . . .

"Differences:

"1. The first Difference I observed betwixt a live Coal and a shining Wood is, that whereas the Light of the former is readily extinguishable by Compression (as is obvious in the Practice of suddenly extinguishing a piece of Coal by treading upon it), I could not find that such a Compression as I could conveniently give without losing sight of its operation, would put out, or much injure the Light, even of small Fragments of shining Wood. . . .

"2. The next Unlikeness to be taken notice of betwixt rotten Wood and a kindled Coal is, that the latter will, in a very few Minutes, be totally extinguished by the withdrawing of the Air; whereas a Piece of shining Wood, being eclipsed by the Absence of the Air, and kept so for a Time, will immediately recover its Light if the Air be let in upon it again within half an hour after it was first withdrawn. . . .

"3. The next Difference to be mentioned is, that a live Coal, being put into a small close Glass, will not continue to burn for very many Minutes; but a Piece of shining Wood will continue to shine for some whole Days. . . .

"4. A fourth Difference may be this: that whereas a Coal, as it burns, sends forth Store of Smoke or Exhalations, luminous Wood does not so.

"5. A fifth, flowing from the former, is, that whereas a Coal in shining wastes itself at a great Rate, shining Wood does not. . . .

"6. The last Difference I shall take notice of betwixt the bodies hereto compared is, that a quick Coal is actually and vehemently hot; whereas I have not observed shining Wood to be so much as sensibly lukewarm."

The light actually comes from tiny threads of cells, the mycelium of a fungus which spreads through the wood as mould on bread. At times the threads form a fruiting body or mushroom, the fungus proper, which may itself be luminescent (Fig. 17). Only if these fruiting bodies form can the fungus be identified. Molish (1904, 1912), to whom much of our knowledge of luminous fungi is due, grew a form which he called *Mycelium X* for many years, without the appearance of a fruiting body.

The culture of fungi is quite easy. The mycelium can often be grown from spores or from dried wood, if not too old, on agar culture media, like luminous bacteria (Fig. 18). Indeed the luminescence of true fungi is quite similar to that of bacteria, continuous and independent of stimulation, although Bose (1930) found sunlight to stimulate the luminescence of an Indian wood in which the light had disappeared from age.

Bacteria actually form one group of the fungi, the schizomycetes, or fission fungi, but they are never responsible for the luminescence of wood. Of the many other groups of fungi, the moulds, yeasts, slime moulds, rusts, smuts, stink horns, puff balls and mushrooms, the latter, known as the hymenomycetes, contain the great majority of luminous species. Such genera as *Agaricus, Armillaria, Pleurotus, Panus, Mycena, Omphalia, Locellina, Marasmius, Clytocybe, Corticum, Fomes, Polyporus, Collybia, Tricholoma, Kalchbrennera* and *Trametes* are said to light.

One of the ascomycetes, *Xylaria hypoxylon*, the stag-horn fungus (Fig. 20), has been observed to luminesce by Ludwig (1874), Crie (1881) and Gueguen (1907), but not by Molish (1904). Possibly this form exists in two varieties, one luminous, the other not. This situation is known for *Panus stipticus*. The European variety is not luminous while the American form is luminescent (Buller, 1924). Macrae (1937) has found that pairings of mycelia, each from a single spore of the American and of the European forms, are fertile, and the resulting diploid mycelium is luminous, i.e. luminescence is dominant. *Ileodictyon (Clathrus) cibarius* and *Dictyophora phalloides* of the gasteromycetes are also said to be luminous. Australia is especially rich in luminous fungi which have been studied by McAlpine (1909).

Although the ancients recorded the existence of luminous wood and many observers since then (Boyle, 1667; Spallanzani, 1796, 1799; A. von Humbolt, 1796, 1799; Carradori, 1797; Gaertner, 1799; Broeckmann, 1801; Heinrich, 1815) have studied the glow, it is not surprising that early controversy largely dealt with the question as to whether the light was from the wood or from something growing on it. Fungal growths were called *Rhizomorpha* and the observations of Bischof (1823), Derschau (1823), Nees von Esenbeck, Bischof and

LIGHT-PRODUCING ORGANISMS

Noggerath (1823), Schmitz (1843) and Heller (1853) established the fact that the *Rhizomorpha* actually produced the light. Heller is usually credited with proving the fungal origin of glowing wood, for he examined the mycelium with a microscope. Fabre (1855) made an extensive study of the respiration of a luminous fungus.

One of the commonest luminous mycelia is that of the "Hallimasch," or honey fungus, *Armillaria mellea*, whose fruiting body does not light. The mycelium of *Clitocybe* is not luminous (Hanna, 1938) but the light is brightest on the gills of the mushroom, and disappears when the fungus is old. Dahlgren (1916) has figured the luminous cells. In *Pleurotus*, the mycelium lights first, then the whole fruiting body, then only the gills. In a *Mycena* from Java, the cap is described as covered by a thick luminous slime.

Dried fungi will luminesce when again moistened but not if first ground into a powder. Neither Kawamura (1915) nor Buller (1924) could obtain a luminous fluid by pressing out the juice of the fresh fungus.

Bothe (1928, 1930, 1931, 1935) has studied the heredity and the conditions of culture of various fungi, especially of the genus *Mycena*, in which some species are luminous, others not. Fungi of this genus frequently live on moulding leaves. Luminous fungi will live on many kinds of nutrient substances, wood, bread or vegetables. Luminous cheese was described by Hermbstädt (1808) and luminous peat by Plot (1686). Buller (1924) reports that almost all samples of old moulding leaves will contain some luminescent mycelium and has described (1934) the luminous leaf-spot fungus, *Omphalia flavida*, which attacks living coffee tree leaves (Fig. 19). The records of luminous potatoes are of interest, especially one in which the officer in charge thought the barracks at Strasburg were on fire as a result of the light emitted by a cellarful

LIVING LIGHT

of potatoes on which a luminous mycelium was growing. In mines especially, the wooden supports of tunnels are noted for their brilliant phosphorescence (Fig. 11).

Luminous wood must not be confused with the "sparkling" of wood. This is due to minute collembolid insects, such as *Neanura muscorum*, living in the decaying material.

LUMINOUS ANIMALS

Authorities differ in the classification of the Animal Kingdom. Some recognize 12 great divisions or phyla, others as many as 32. Most difference of opinion regarding phyla has been over small aberrant groups with only a few species. If we accept 17 great phyla in the Animal Kingdom, the following table gives the approximate number of species in each.[4] Those with luminous forms are italicized. Of the 17 phyla, 11 contain luminous forms and one is doubtful; of 49 classes of animals 21 contain luminous forms and one is doubtful; of 234 orders of animals 39 contain luminous forms and four are doubtful.

PHYLUM	NO. OF SPECIES
1. *Protozoa* (unicellular forms)	15,000
2. *Porifera* (sponges)	3,000
3. *Coelenterata* (jellyfish, corals, sea anemones)	9,500
4. *Ctenophora* (comb jellies)	100
5. Platyhelminthes (flatworms)	6,000
6. *Nemertea* (proboscis worms)	500
7. Nemathelminthes (thread worms)	3,000
8. Trochelminthes (wheel animalcules)	1,200
9. *Bryozoa* (moss animals)	3,000
10. Brachiopoda (lamp-shells)	120
11. Phoronidea (phoronids)	15

[4] From Pratt's *Manual of the Common Invertebrate Animals* (1935).

LIGHT-PRODUCING ORGANISMS

12. *Chaetognatha* (arrow-worms)   30
13. *Annelida* (earthworms and marine worms)   6,500
14. *Arthropoda* (insects, crabs, spiders, centipedes) 640,000
15. *Mollusca* (snails, squid, shellfish)   70,000
16. *Echinodermata* (sea urchins, starfish)   4,800
17. *Chordata* (ascidians, vertebrates)   60,000

PROTOZOA

This group contains four classes, but only the flagellates and radiolarians are luminous.

Flagellates. Many of these microscopic organisms are claimed by botanists as well as zoologists. The dinoflagellates may not only contain chlorophyll or other pigments and be covered by cellulose plates, plant characteristics, but they may also ingest food as do animals. Kofoid believes that practically all marine species are luminescent, giving rise to the extraordinary displays of phosphorescence in the sea described by so many voyagers. No luminous fresh-water flagellates have yet been described.

Biornonius (1674) speaking of Iceland says, "*Our seawater in clear Nights, being struck with Oars, shineth like Fire bursting out of a Furnace.*" Both cold and warm waters may luminesce.

Darwin in the *Voyage of the Beagle* (p. 176) in 1832 refers to a brilliant display in the South Atlantic: "While sailing a little south of the Plata on one very dark night, the sea presented a wonderful and most beautiful spectacle. There was a fresh breeze, and every part of the surface, which during the day is seen as foam, now glowed with a pale light. The vessel drove before her bows two billows of liquid phosphorus, and in her wake she was followed by a milky train. As far as the eye reached, the crest of every wave was bright, and the sky above the horizon, from the reflected glare of these livid flames, was not so utterly obscure as over the vault of the heavens."

Noctiluca miliaris is a name so well known that many persons cry, "Noctiluca!" when the fiery wake of a boat is seen at night. Many are the exclamations of surprise as a ship ploughs through a soup of these organisms, which often color the water pink by day because of a red pigment they contain. The smaller dinoflagellates are also red or yellow when in large numbers and the Red Sea derives its name from such accumulations.

Noctiluca is a slightly indented sphere, one millimeter in diameter, just visible to the naked eye (Fig. 21). Because of its size, it played a large part in the gradual recognition that all phosphorescence of the sea was due to living creatures and the history of its discovery forms an interesting chapter of animal luminescence. It is sometimes said that Dartous de Mairan, in his prize essay published in 1717, *Dissertation sur la Cause de la Lumière des Phosphores et des Noctiluques*, refers to the organism, *Noctiluca*, the cause of much of the light of the sea. This is not true. The word "noctiluca" was then used for any kind of phosphorescence and Robert Boyle refers to the Bolognian phosphor (impure barium sulphide) as a solid noctiluca. He also speaks of gummous, liquid and aerial noctilucas, meaning the element phosphorus in solid, dissolved and gaseous form.

Baker, in the first edition of his book (1753), *Employment for the Microscope*, in a chapter "On luminous water insects" probably refers to the animal, *Noctiluca*, described as little globes with tiny tails, in a letter to him from Joseph Sparshall. A figure accompanied the letter but came too late to be reproduced.

Le Roi (1754), Baster (1757) and Le Gentil (1761) probably also saw *Noctiluca* as they speak of luminous particles like the head of a pin or tiny globes like fish eggs, while Rigaud (1765), a physician at Calais, certainly saw the animal.

In 1775 Dicquemare figured *Noctiluca* in recognizable form for the first time, and Slabber, a Dutchman and friend of Baster, in 1781, also illustrates it perfectly as *Medusa marina*. Macartney (1810), from observations first made in 1804, called it *Medusa scintillans* and gives a small scale figure. All three are reproduced in Fig. 22.

In 1816 Suriray, a physician at Havre, wrote a letter to Lamarck describing the organism, which Lamarck included in his *Systeme des Animaux sans Vertebres* as *Noctiluca miliaris*, Suriray. According to Kofoid (1921) who removed *Noctiluca* from the old order of cystoflagellates, it should be called *Noctiluca scintillans*. Thus ends the history of a famous organism.

Pratje (1921) has collected the various observations on *Noctiluca* in an exhaustive morphological and physiological study. He finds 12 per cent of lipoid material in the animal, higher than in most protozoa. The intracellular luminescence appears only on stimulation and comes from small granules at the cell periphery and scattered through the protoplasmic strands so carefully described and figured by Quatrefages (1850), who noticed that the glow frequently begins in two spots, one on each side of the oral groove (Fig. 6). These move or spread over the *Noctiluca* as a wave, but the actual points of light are stationary, lighting and then going out. Details of the physiology of stimulation will be described in Chapter V.

Many dinoflagellates are truly microscopic, one of them only 16 μ long. The genera *Ceratium*, *Peridinium*, *Prorocentrum*, *Pyrodinium*, *Gonyaulax*, *Blepharocysta*, *Amphidinium*, *Pyrocystis*, *Cochlodinium* and *Gymnodinium* are known to be luminous and there are many others. The Fire Lake near Nassau in the Bahamas was for many years famous for its content of one form, *Peridinium bahamense*, described by Plate in 1906 (Fig. 23). In the Pacific, Kofoid (1911) finds that the great displays of luminescence are due to *Gonyaulax polyhedra* (Fig. 26), while Dahlgren (1915) observed a luminous colonial form in Chesapeake Bay. Any fresh sea water filtered through cloth at night shows minute transient points of light which are most likely to be these forms. Michaelis (1830) and Ehrenberg (1831, 1834, 1859) described many species while Kofoid and Swezy (1921) have written a monograph on the unarmored dinoflagellates. One of their plates is reproduced as Fig. 24.

Reinke (1898) showed that these smaller forms also light only on stimulation and Zacharias (1905) studied the effect of many chemicals. A method of artificial cultivation of luminous flagellates is urgently needed, as these forms would be superb material for luminescence studies if they were not so capricious in occurrence.

Radiolaria. As the name indicates, these animals are radially symmetrical single cells living either alone or as colonies in a mass of jelly. They would only be noted as luminous by careful observation and Tilesius (1819) is said to be the first to have observed their luminosity. Baird (1830) and Mayen (1834) also saw them light, while Brandt (1885) made careful studies of their luminescence and structure. The genera, *Thalassicola*, *Myxosphaera*, *Collosphaera*, *Sphaerozoum* and *Collozoum* have been described as luminous but all radiolarians may possess this power.

The author has observed radiolaria at Naples and noted particularly that the rather bluish intracellular light, which appears only on stimulation, is diffuse and weak, not the bright flash or sparkle characteristic of the flagellates. Each cell is made up of a perforated capsule, within which are one or many nuclei with a central or several "oil" drops. Outside the capsule

LIGHT-PRODUCING ORGANISMS

are yellow alga cells scattered about the ectoplasm (Fig. 25). There has been some discussion as to whether the oil is the source of the light, for Brandt observed that it is the center of the individual that luminesces. Perhaps the protoplasm next to the oil contains minute photogenic granules.

PORIFERA

The sponges or pore bearers are hard to classify. They have been regarded as plants, as animals like polyps, and as colonial flagellate protozoa. Their fundamentally animal nature is certain but their characteristics are so unusual that they are sometimes contrasted with all other many celled animals as the division *Parazoa*, a side issue.

There is every reason to predict that luminous sponges should occur since they are surrounded by groups in which luminescence is widespread. A number of early accounts are doubtful, since so many creatures which are luminous may live in the tissues of the sponge. A clear case was described by Okada (1925), in the sponge, *Crateromorpha meyeri*, from the bottom, 1,000 meters deep in the Sagami Sea. Bluish starry spots appeared on dredged specimens which became especially bright when the sponge was placed in fresh water. They turned out to be worms of the alciopinid group and the sponge was not self-luminous. Dahlgren (1916) and Trojan (1933) have also observed luminous animals living on a sponge. The author (1921) described a *Grantia* from Friday Harbor, Wash., in which the character of the luminescence was so similar to that of coelenterates that it was classed as self-luminous. The yellowish light appeared from every specimen when it was rubbed with the hand and was especially bright on crushing. Squeezed through cheesecloth a long lasting luminous extract was obtained, containing points of light which appeared and disappeared exactly like those in extracts of coelenterates. Addition of fresh water or saponin resulted in a great increase in the light just as in extracts whose self-luminosity is unquestioned. No dinoflagellates or other luminous animals were visible.

COELENTERATA

Hydroids, jellyfish, sea pens, sea fans and siphonophores are all closely related animals belonging to the great group of coelenterates, the simple two-layered animals whose body cavity and digestive cavity (*enteron*) are the same. Coelenterates, which include also sea anemones and corals, none of which are luminous, are plant-like, but no careful observer could mistake them for anything but primitive multicellular animals. Light appears to be linked with life, for the simplest organisms we know of, bacteria, fungi, protozoa, sponges and coelenterates are often luminescent. The living world is weighted at the very bottom with light-producing forms.

Hydroids and jellyfish may be different forms of the same species. They show the interesting alternation of generations, a medusa or jellyfish generation giving rise to a hydroid generation and hydroids producing jellyfish again. In some forms of jellyfish the hydroid stage is rudimentary, while some jellyfish produce only rudimentary hydroids. They have been classified as both hydroids and medusae and older workers did not realize they were different aspects of the same animal. Less is known of the luminescence of hydroids than of medusae, but *Aglaophenia, Campanularia, Sertularia, Gonothyrea, Obelia* and *Clytia* have been described as genera containing some luminous species (Fig. 27). Darwin (1839) saw, near Tierra del Fuego, a species allied to *Clytia* luminesce and noted that the light always travelled up the branches from the base to the extremities.

LIGHT-PRODUCING ORGANISMS

Luminous medusae were known to Pliny and must have been observed by many persons, for the large Mediterranean form, *Pelagia noctiluca*, is a striking object at night. The allied genera *Charybdea*, *Chrysaora*, *Cyanea* and *Rhizostoma*, as well as the less closely related *Stomatoca*, *Clavula*, *Rathkea*, *Laodicea*, *Phialidium*, *Halistaura*, *Siriope*, *Solinissus* and *Aequorea* contain some luminous species. Early descriptions of luminous medusae are given by Gesner (1555), Kircher (1646), Loffling (1758), Slabber (1778) and Spallanzani (1794) but it was not until 1841 that Hassal described the luminescence of hydroids. Panceri (1872) made a number of interesting observations on both hydroids and medusae.

In *Pelagia* (Fig. 31) the whole of the outer surface of the animal, including tentacles, is luminescent on stimulation, a gentle touch causing a local light spot which spreads if the stimulation is more violent. The spread of luminescence is due to a nerve net similar to that involved in the control of pulsations of the animal. When handled, the luminous material sticks to the hand, a luminous mucus which can be seen to glow in discrete points under the microscope. These luminous granules, whose breakdown is accompanied by a burst of light, are particularly well seen when the slime is treated with fresh water or with well known cytolytic agents such as diluted sea water, saponin or sodium glycocholate.

Other jellyfish contain luminescent spots, sometimes on the umbrella, sometimes on the tentacles. At Friday Harbor, Wash., a small form, *Aequorea forskalea*, 1½ to 3 inches in diameter, is common. It has yellow masses of photogenic cells around the rim of the umbrella at the base of the tentacles. On merely touching a jellyfish or stimulating it electrically, one cannot observe that any luminous secretion is definitely thrown into the water, but on gentle stroking of the edge of the umbrella a mass of luminous material comes off which adheres to the fingers, and on tossing an animal on the surface of the water, abundant luminous material is liberated that causes the sea water to glow. This material can be dried and will give a bright light when again moistened. Just as in *Pelagia*, the slime is full of granules which dissolve with the emission of light giving under the microscope an appearance of a starry sky (Harvey, 1921).

Luminous siphonophores were noted by Panceri (1872) and Giglioni (1870). The genera *Praya*, *Hippopodius*, *Diphyes*, *Abylopsis* and *Algama* contain luminous species. The author has seen the delicate *Algama elegans*, common at the mouth of the Bay of Fundy, luminesce with a faint bluish light when stimulated.

Sea fans or gorgonians, dredged from considerable depths, were observed to glow with a pale lilac light by Sir Wyville Thomson on the *Challenger* expedition; also alcyonarians and antipatharians. It is sometimes difficult to identify the species referred to by older observers and a careful survey of these groups is much needed. The common shallow-water gorgonians of Florida reefs are not luminous.

Pennatulids or sea pens (Fig. 29) were known to be luminous since earliest times and have been described by Gesner (1555), Ellis (1763), Imperati (1672), Spallanzani (1796), Tilesius (1819) and Grant (1827). The genera *Pennatula*, *Veretillum*, *Cavernularia*, *Funiculina*, *Umbellula*, *Leioptilus* (=*Ptylosarcus*) and *Pteroeides* contain luminous species.

Sea pens, unlike medusae, represent colonies of animals. On stimulation, a wave of luminescence passes along the stem from one to another of the individuals (zooids) making up the colony. This was carefully described by Panceri (1872) who made many experiments on the Naples form, *Pennatula phosphorea*, recently studied by Moore (1926). Ellis described *Renilla* in 1763 and Louis Agassiz observed in

1850 that this "sea pansy," "shines at night with a golden green light of a most wonderful softness." Parker (1920) described the light as "clear blue green" spreading from a spot stimulated in "luminous ripples—which spread out concentrically over its surface like waves on the smooth face of a pond into which a pebble has been thrown."

Renilla is a kidney-shaped colony of zooids on a stalk, so closely united by nerve and muscle connections that a local stimulus sets up not only the luminous waves but also impulses which cause retraction of the zooids (a protective reaction) (Fig. 28). The luminescence comes from small masses of light colored material that stud the upper surface of the colony at the base of the zooids. No light is visible from stalk or undersurface but impulses which stimulate luminescence pass over these non-luminous regions. In fact the colony can be cut into any form or shape and the impulses will follow the intact surfaces but they do not pass across a cut. As Parker (1920) has pointed out, *Renilla* behaves more like an individual than a colony.

The author (1917) has studied *Cavernularia haberi*, a large pennatudid which lies hidden in the sand at the bottom of fjords in Japan during the day but at night takes up water and expands. Large colonies may extend to a length of two feet. When stimulated electrically or by touching or by addition of ammonia to the sea water, a brightly luminous slime is formed. The whole of the outer surface of the colony can form the luminous slime but not the spongy inner material. The stalk containing no zooids is especially brilliant. Both medusae and pennatulids appear to exemplify extracellular luminescence, although Panceri believed the light was intracellular and that rubbing the animal scraped off some of the luminous epithelium. Sections of the luminous surface of *Pelagia* have been figured by Dahlgren (1916). They are typical of forms producing a luminous secretion and show single mucous and luminous gland cells.

CTENOPHORA

Commonly called comb bearers or comb jellies, these animals are transparent masses of jelly whose slow locomotion through the water is not by pulsation but by means of eight rows of paddle plates which look like the teeth of a comb (Fig. 30). These plates wave in the water liked fused cilia. Probably all species are luminescent but only in darkness. Daylight or bright artificial light prevents the luminescence, which returns after some time in the dark. They have been described by Dicquemare (1775), Hulme (1803), Tilesius (1819) and many others. Such genera as *Pleurobrachia, Mnemiopsis, Bolinopsis, Cestus, Eucharis* and *Beroë* are common. These forms may be very abundant in the sea and give rise to the larger discrete diffuse flashes of light observed along the side of a ship at night.

The light comes from cells, studied by Dahlgren (1916), along the canals under the rows of paddle plates—brilliant greenish streaks (Fig. 32) when the animal is touched with a glass rod. No luminous secretion is observed on the outside of the animal but material might be secreted into the canals. In a Japanese form, *Ocyropsis fusca*, Okada (1926) has described luminescence in prolongations (lappets) of the animal without paddle plates. The light always appeared in regions where the sexual cells are formed, which are possibly luminescent. Unsegmented eggs of *Mnemiopsis leidyi* are not luminous, but early segmentation stages are luminous (Allman, 1862; Agassiz, 1874; Peters, 1905). Yatsu (1912) observed at Naples that eggs of *Beroë* gave "a beautifully greenish light" when stimulated electrically. It was restricted to the ectoplasm and no visible granules could be seen. If ctenophores

LIGHT-PRODUCING ORGANISMS

are gently rubbed with the fingers, no luminous slime appears, since the luminous cells are within the canals, but if shaken or squeezed through cheesecloth they yield a luminous extract whose behavior is like that of medusae and pennatulids already described. The general similarity of light production in all coelenterates and ctenophores is clearly apparent. The ctenophores differ from other coelenterates chiefly in that their luminescence is suppressed by light in a manner that will be fully described in Chapter V. Ctenophores, like other forms, are fatigued on repeated stimulation and luminesce much less strongly, but recover their power after rest.

PLATYHELMINTHES

This phylum contains the flatworms and tapeworms, many of them parasitic. The marine turbellarian worm, *Typhlophana retusa* was described as luminous by Viviani (1805) but no recent confirmations have been made.

NEMATHELMINTHES

No luminous forms are known in this phylum which contains the roundworms, threadworms, and spiny headed worms, mostly parasites.

NEMERTEA

This phylum of worms is named from a Greek root meaning "unerring" since their probosces, which can be thrust out quickly, never miss their prey. No luminous nemertean had been found until Kanda (1939) discovered one, identified by Kato (1939) as *Emplectonema kandai*. It lives coiled up on an ascidian, *Chelyosoma siboja*, which is common in Aomori Bay, 35-40 meters deep. The luminescence is described by Kanda as appearing over the whole body except the tip of the head, when the animal is handled or strongly stimulated. As this form may be a meter long and 0.5 to 0.7 mm. in diameter when stretched, the appearance of a sinuous greenish line of fire must be an extraordinary sight. Touched locally with a glass rod the light does not spread far, but on stretching, the whole worm lights. No slime comes off on the fingers. Dried animals can produce light when again moistened but luciferin and luciferase cannot be demonstrated. Two kinds of cells are found in the surface of the body, mucous and photogenic cells. They appear to have ducts opening through the cuticle and a small amount of material must be secreted but not enough to stick to the fingers. There is no evidence that the light is bacterial in origin.

TROCHELMINTHES

This phylum of wheel worms include the rotifers and two small groups of minute animals, gastrotrichs (*Chaetonotus*) and kinorhynchs (*Echinoderes*) neither of which are luminous.

The marine rotifer, *Synchaeta baltica*, was described as luminous by Michaelis (1830), but Ehrenberg (1834) could not confirm this observation. He gives a detailed figure of the animal. The light might be due to bacterial infection or the dinoflagellate food eaten. We have no certain record of any self-luminous rotifers.

BRYOZOA

Moss animals are so called from their habit of covering rocks, shells and seaweed. They are divided into the endoprocts and ectoprocts. Only the latter may contain luminous forms although the reports need confirmation and nothing is known of the photogenic cells. *Membranipora pilosa* = *Electra pilosa* (Molish, 1912), *Flustra membranacea* = *Membranipora membranacea* (Landsborough, 1842), *Retepora* (Ehrenberg, 1934) and *Scrupocellaria reptans* (Gadeau de Kerville, 1890) are said to light.

LIGHT-PRODUCING ORGANISMS

BRACHIOPODA

No member of this small group of lamp-shells is definitely known to be luminous.

PHORONIDEA

These small sessile marine worms, of which only 15 species have been described, are not luminous.

CHAETOGNATHA

This phylum contains *Sagitta*, a transparent arrow-like organism and a common pelagic form in tropical seas, the type of animal which might be expected to luminesce. The reports that they do are due to Giglioni (1870) but have not been confirmed by later observers. The author has never seen a luminous *Sagitta* and we can only say that luminescence in these animals is doubtful but possible.

ANNELIDA

The great group of ringed or segmented worms is divided into a number of groups. Of these only the oligochaetes, mostly earthworms with few setae, and the polychaetes, mostly marine with many setae, are luminous.

Earthworms. Luminous forms were observed by Grimm (1682), Flaugergues (1771), Brugiere (1792) and numerous others since then from every part of the world. They are described as shining with a nearly white light resembling "a bar of iron heated to white heat." In some the light is said to appear at the coupling season and to be most marked about the clitellum, disappearing immediately after copulation. Other forms are certainly luminous the year round. The genera *Eisenia, Microscolex, Enchytraeus, Octochaetus, Eutryphoeus, Megascolex, Henlea* and *Chilota* are luminous. All observers have noted that a slime is shot out by the animal, when irritated, that may cover the ground, leaving quite a trail of light. Gilchrist (1919) studied a South African species (*Chilota*) and Skowron (1926) has investigated *Microscolex phosphorea* with special attention to the occurrence of symbiotic organisms in the slime, since Pierantoni (1924) has claimed that the light in the latter form is due to symbiotic bacteria. The fact that earthworms live in damp decaying soil would make it probable that infection from bacteria or fungi might occur.

In both *Chilota* and *Microscolex* it can be seen under the binocular microscope that luminous slime comes mostly from the anal opening, occasionally from the mouth. In other worms, such as *Eutyphoeus*, *Megascolex* and *Eisenia*, Gates (1925) and Komarek (1924) have described it as coming from dorsal pores along the whole of the body. In *Microscolex* a cut through the body wall but not as deep as the intestine strikes luminous material which is formed by cells in the coelom. This passes into the intestine near mouth and anus by way of the nephridia, to be ejected by the animal. In dying or dead animals a glow may come from inside the body. In the slime are found round cells full of granules and granules themselves. If dried, light will appear again when moistened. Centrifuged granules luminesce when treated with ether, but a filtrate through a silicious filter does not. Light comes from the dissolving liberated granules whose behavior when treated with various reagents is so different from that of any known bacteria that Skowron (1928) believes they cannot be considered as such.

Marine worms. The genera of luminous marine worms are found in eight groups: Polynoid (*Acholoë*, *Polynoë*, *Harmothoë*); Alciopid (*Corynocephalus*); Tomopterid (*Tomopteris*); Syllid (*Eusyllis*, *Odontosyllis*, *Syllis*); Nereid (*Perinereis*, *Nereis*); Cirratulid (*Cirratulus*, *Heterocirrus*, *Macro-*

LIGHT-PRODUCING ORGANISMS

chaeta); Chaetopterid (*Mesochaetopterus, Chaetopterus*) and Terebellid (*Polycirrus, Thelepus*).

Some show more or less general luminescence from various parts of the body, while in others the luminescence is restricted to definite spots or organs, a localization that becomes more and more common in the higher forms. Early observers are de la Voie (1666) who noted worms (probably terebellid) on oyster shells and Vianelli (1749), Griselini (1750) and Nollet (1750) who saw luminous worms in the canals of Venice. At this time such forms were spoken of as *Scolopendra marina*. Linnaeus (1758) called a luminous worm, *Noctiluca marina*.

Chaetopterus is one of the most striking of luminous animals, first described as luminous by Will (1844). Almost the whole worm glows when disturbed in any way, the brightest luminous areas varying somewhat in different species (Fig. 33). Even the larva is luminous (Enders, 1909). As the animal spends its whole life in a parchment tube buried in the sand, only the two ends protruding, inlet and outlet for a supply of fresh sea water, the use of the light is a puzzle. Dahlgren (1916) has portrayed the worm shining brightly as it is attacked by an eel (Fig. 34).

The secretion of photogenic cells scattered throughout the skin of the animal forms a luminous slime that adheres to the fingers. Careful histological studies have been made by Enders (1909), Trojan (1913, 1919), Dahlgren (1916), Krekel (1920) and Fujiwara (1925). In the epithelium are found photogenic cells with rather large granules, together with mucous cells (Fig. 36). The former are probably under nervous control since a wave of light passes over the animal when the head end is stimulated.

Terebellid and cirratulid worms have a huge mass of tentacles which secrete a luminous slime like that of *Chaetopterus*. Sometimes the body is luminous also. A small form often lives on oyster shells (Fig. 37), while a large one is common at Plymouth, England, living under stones. The secretion of luminous slime is quite analogous to that of *Chaetopterus* (Dahlgren, 1916).

The outstanding syllid worm is *Odontosyllis enopla*, the fireworm of Bermuda, observed by many and studied by Galloway and Welch (1911) and Dahlgren (1916). A closely related form (*O. phosphorea*) is found near Vancouver, B.C. (Potts, 1913; Frazer, 1915). The light is here a definite mating signal. The Bermuda worms, which live in coral at the bottom, show a lunar periodicity, swarming the second, third and fourth days after full moon. E. L. Mark has kept records for many years. They appear at the surface on these days, 55 minutes after sunset throughout the year, whether the sky is overcast or clear, whether winter or summer. The swarming lasts for 15 to 30 minutes and then ceases abruptly. The author can vouch for the account of mating behavior which Galloway and Welch have described as follows:

". . . Both sexes are distinctly phosphorescent: the female with strong and more continuous glow, and the male with sharper intermittent flashes.

"In mating, the females, which are clearly swimming at the surface of the water before they begin to be phosphorescent, show first as a dim glow. Quite suddenly she becomes acutely phosphorescent, particularly in the posterior three-fourths of the body, although all the segments seem to be luminous in some degree. At this phase she swims rapidly through the water in small, luminous circles two or more inches in diameter. Around this smaller vivid circle is a halo of phosphorescence, growing dimmer peripherally. This halo of phosphorescence is possibly caused by the escaping eggs, together with whatever body fluids accompany them. At any rate the phosphorescent effect closely accompanies ovulation, and the eggs continue mildly phosphorescent for a while." . . .

"The male appears first as a delicate glint of light, possibly as much as 10 or 15 feet from the luminous female. They do not swim at the surface, as do the females, but come obliquely up from the deeper water. They dart directly for the center of the luminous circle and they locate the female with remarkable precision, when she is in the acute stage of phosphorescence. If, however, she ceases to be actively phosphorescent before he covers the distance, he is uncertain and apparently ceases swimming, as he certainly ceases being luminous, until she becomes phosphorescent again. When her position becomes defined he quickly approaches her, and they rotate together in somewhat wider circles, scattering eggs and sperm in the water. The period is somewhat longer on the average than when the female is rotating alone; but it, too, is of short duration.

"So far as could be observed, the phosphorescent display is not repeated by either individual after mating. Very shortly the worms cease to be luminous and are lost. . . ."

Galloway and Welch thought the luminescence came from twisted gland cells in the skin but Dahlgren has described from sections a large gland with a duct on the lower part of a parapodium. This may be a luminous gland but the source of the light has never been studied in living animals and it is possible that cells from the body cavity are involved, as in earthworms.

Tomopteris (Fig. 37) is an extraordinary transparent pelagic worm with groups of luminescent cells on its parapods. The author remembers the light of a form collected at Plymouth, England, as yellowish. Vejdovsky (1878) first described the organs as eyes, while Greef (1882), Kiernik (1908) and Meyer (1929) saw their luminescence.

The organ represents a transformed nephridial funnel and consists of a rosette of photogenic yellow cells with a melon-shaped mass of transparent cells nearly surrounding it, and a nerve ganglion at its base. Luminescence is presumably intracellular.

In the polynoëid worms the luminous organs are on overlapping scales which contain nerves and receive a blood supply. They cover the back of the worm. A scale removed from the body can be stimulated to luminesce in minute points of greenish blue light that comes from papillae most abundant on the edge of the scale. No light can be rubbed off on the fingers although the luminous gland has a duct. Stimulation and luminescence have been studied by Jourdan (1885, 1887), Falger (1908), Kutchera (1909) and Dahlgren (1916). The animal lights progressively tailward when stimulated on any part of the body. The light papilla is supplied by a nerve and is evidently under the control of the animal. Many a salt-water tragedy must occur at night like that shown in Fig. 40, in which the light of the hind end of the worm is bright but the front half is dark and crawls away, possibly to regenerate a new tail. The luminescence may be said to protect the animal as a "sacrifice lure." Acholoë (Fig. 38) lives with starfish.

Luminous alciopid worms were found by Okada (1925) living in a sponge while nereids have been reported as luminous by the older workers.

ARTHROPODA

This largest phylum of the animal kingdom contains at least 640,000 species divided among six classes. These are the (1) crustaceans, (2) myriapods, (3) insects; (4) onychophores, with only one genus, *Peripatus*, a non-luminous, worm-like

LIGHT-PRODUCING ORGANISMS

arthropod; (5) the king crab, Limulus, a relic of the past, and (6) spiders. Some spiders have been reported to be luminous and the first three classes contain undoubted luminous representatives.

Crustaceans. In this class five groups luminesce, the mysids, ostracods, copepods, schizopods and decapods. Some secrete from glands an abundant luminous fluid into the sea, others possess complicated photophoros with lens and reflector, others both types of organ.

It is somewhat difficult to tell just what animal the older observers referred to in their description of luminous crustacea. Kircher (1640) speaks of a luminous crab but this may have been infected with luminous bacteria, since there are no self-luminous crabs. Anderson (1747) described *Oniscus fulgens*, Banks (1768) *Cancer fulgens*, both evidently shrimp. Fabricious (1780) described *Cyclops brevicornis*, which may have been a copepod.

Godeheu de Riville (1754) undoubtedly observed luminous ostracods, among which the form, *Cypridina hilgendorfii*, has been most useful for chemical work. The genera *Pyrocypris*, *Gigantocypris* and *Conchoëcia* also contain luminous species. A very large gland near the mouth supplies a luminous secretion which can be treated by the ordinary chemical methods of manipulation. *Cypridina*, only 1/8 inch long, is called "marine firefly" (*Umihotaru*) by the Japanese fishermen, who catch them as they feed on fish heads placed in the sea as bait. Other species may be collected by towing in the open ocean. The quickly dried animal retains its power to luminesce for years whenever moistened. It is this dried material which has been used for the chemical work described in Chapter IV.

The animal itself is covered by a hinged chitinous shell that almost completely hides the swimming legs. Dried organisms look like seed (Fig. 41). In nature a touch of the living animal will stimulate emission of the brightly bluish luminous fire. The structure of the glands has been described by Watanabe (1897), Doflein (1906), Dahlgren (1917), Yatsu (1917), Okada (1926) and Takagi (1936). Two kinds of cells can be easily seen in the living material, one full of large yellow granules, the luciferin, the other filled with small colorless granules, the luciferase. Muscles squeeze out these granules. On meeting in the water they dissolve and emit light in a manner to be described in detail in Chapter IV. Four types of gland cells can be made out in fixed sections (Fig. 35) and are figured by Dahlgren, Okada and Takagi.

Copepods are widely distributed in arctic and temperate seas. The genera, *Metridia, Pleuromma, Lucicutia, Heterorhabus, Onchaea, Pontella, Chiridius, Euchaeta* and *Corycaeus* contain luminous species. Naples is a favorable place to study many of these forms and Giesbrecht (1895) has given an excellent account. The light material is found in groups of secretory cells of a greenish color on head, back or tail, sometimes on the leg (Fig. 39). The secretion is expelled when the animal is disturbed but never in large enough amounts to surround the animal with a luminous fire. Only in the late winter and spring is the lighting power apparent. Even the nauplius stages of these copepods have similar light glands. The luminous secretion is said to be granular or like drops of a greenish oil. In ostracods and copepods both sexes are luminous and both have the same distribution of luminous glands. Moreover many of the luminous species have no eyes so that the function of the light would appear not to be a sex attraction and its use is problematical.

Schizopods and decapods, containing the shrimp and the prawns, often look alike. A few are fresh-water inhabitants, whose luminescence by infection with luminous bacteria has already been described for the decapod, *Xiphocarida*, of Lake

LIGHT-PRODUCING ORGANISMS

Suwa, Japan (p. 30). Most are surface-living or deep-sea forms and have been caught at depths up to 4,200 meters. The genera *Plesiopenaeus, Gennadas, Amalopenaeus, Sergestes, Systellaspis, Oplophorus, Acanthephyra, Heterocarpus, Polycheles* and *Thalassocaris* of the decapods contain luminous organs. Among schizopods, *Thysanopoda, Euphasia, Nyctiphanes, Meganyctiphanes, Nematoscelis, Thysanoëssa, Stylocheiron, Gnathophausia* and *Gastrosaccus* are especially well known luminous genera. The genus, *Mysis*, is said to contain luminous species but some allied mysids such as *Siriella* have probably been mistaken for that genus.

Chace and Nunnemacher (1937) as well as Waterman, Nunnemacher, Chace and Clarke (1939), continuing the older work of Murray (1885), Chun (1890), Brauer (1906, 1908) and Fowler (1905, 1909), have studied the diurnal migration of luminous shrimp. They are rapid swimmers. The deep-sea forms migrate upward at night as daylight fades, and downward in the morning sometimes as much as 200 to 600 meters. In fact they often come to the surface in large numbers at night to breed, in Puget Sound in April, at the mouth of the Bay of Fundy in July and August. In Japan they are caught, dried and sold in the market as a base for soup.

The punctate light organs of many shrimp, photophores, are for intracellular luminescence. They have been carefully described by Vallentin and Cunningham (1888), Chun (1896), Giesbrecht (1896), Hansen (1903), Illig (1905), Trojan (1907), Kemp (1910), Terao (1917) and Dahlgren (1917). They are remarkably like eyes as can be seen from Fig. 42, and were described as accessory eyes until members of the *Challenger* expedition observed their luminescence. They occur on the eye stalks and thoracic and abdominal segments and appendages, sometimes as many as 150. Photogenic cells, lens and reflector can be distinguished. Sometimes the lens is complicated, giving the appearance of an achromatic triplet. Terao has described the luminescence of the photophores of *Sergestes prehensilis* as a dim greenish yellow, lasting for only a moment, and often appearing first in the head region and progressively lighting towards the tail. Dahlgren observed *Nyctiphanes norwegica* giving flashes as bright as those of a firefly. Trojan noted that the larvae of schizopods were luminous.

The pattern of photophores differs in different species and is probably a means of recognition, since the eyes of those species with photophores are larger than the ones without light organs (Chun, 1893, 1896; Welch, 1937, 1938). It has long been recognized that the eyes of many deep-sea animals are unusually large while cave forms have degenerate eyes or lack them, a difference no doubt correlated with the absence of luminescent creatures in caves and their abundance in the depths of the ocean.

The forms which pour out a luminous secretion from glands on the under side of the head are most striking (Fig. 43). Alcock (1903) observed specimens from the depths of the Indian Ocean, and Beebe (1927) in the Atlantic. The author has studied *Acanthephyra pupurea*, caught 1,200 to 1,600 meters deep eight miles south of Bermuda. It will live a long time in cold sea water. This form has a row of black dots along the sides which are luminous organs but no light came from them. When touched with a rod, a cloud of bluish luminescent secretion shot from glands near the mouth and was carried by convection currents throughout the pail, making the sea water luminesce for some time (1931).

Spiders. Luminous spiders (arachnids) have been reported (Brown, 1925) from Burma but not caught. The light might be due to a true light organ, to infection with luminous bacteria or to the eating of fireflies, if the spider could suck in large enough pieces of luminous organ.

LIGHT-PRODUCING ORGANISMS

The luminous "sea spider," *Colossendeis gigas*, a pycnogonid, dredged by Alcock (1902) from the depths of the Indian Ocean, might also be referred to here. It was twenty inches across, greenish blue luminescent on the under side of the legs.

Myriapods. Since these animals, like earthworms, live in decaying leaves and under stones and trunks of fallen trees, many have supposed their light to be an infection from fungi or bacteria. There is, however, no doubt of their self-luminosity and its similarity to that of earthworms. They have been observed by many early workers including Fernandez de Oviedo (1538), Garman (1670), and Ray (1710). Later, Audouin (1840), Brodhurst (1881), Passerini (1882), Richard (1885), Haase (1889), Dubois (1893), Ludwig (1901), Brockhausen (1902) and Arndt (1924) gave descriptions. Millipeds of the family *Xystodesmidae*, the genera *Fontaria* and *Xystocheir*, have been described as luminous (Bruner, 1891) but most of the luminous species are geophilids, long slender centipedes with as many as 173 segments in some species, belonging to the genera *Geophilus*, *Orya*, *Scolioplanes*, *Stigmatogaster*, *Himantarium* and *Orphnaeus*. A luminous slime is secreted from the under surface of the body that sticks to the hands or the surface over which the myriapod crawls. The secretion is squeezed out by muscles from glands in the integument, possibly also from the anus or from preanal glands. The light is said (Gazagnaire 1880-1890) to be seasonal, as in some earthworms, which the secretion resembles, in that many small granules are present, which Dubois (1893) has called "vacuolides" and described as turning into crystals. However, both Brade-Birks (1920), studying *Geophilus carpophagus*, and Koch (1927), in a detailed study of *Scolioplanes crassipes*, noted luminescence the year round and think the purpose is for protection and not sexual attraction, as these species are blind.

Insects. Of the twenty-four orders of insects, only the springtails, flies, moths and beetles appear to contain self-luminous forms. Caterpillars and midges (see pp. 29-30) are known to become infected with luminous bacteria, and reported instances of luminous termite nests (Knab, 1909), mayflies (Hagen, 1873, Eaton, 1882), mole-crickets (Ludwig, 1891) and ants (Wheeler, 1916) may belong in the same category. Branner (1910) gives reports of luminous termites and of fireflies on termite nests but never saw any himself. Possibly the luminous larvae of fireflies live in termite nests. The lantern-fly (*Fulgora*) is not luminous.

Springtails are small wingless insects living in decayed leaves and were first observed by Allman (1850) to be luminescent. Dubois (1886) described them as glowing with a steady light, which increased when disturbed, and when crushed, the entire body seemed to light. The genera *Lipura* (Fig. 44), *Neanura*, *Onychiurus* and *Achorutes* are luminous. The most recent work is by Stammer (1935) and Heidt (1936), who saw the bright green light of *Onychiurus armatus* and *Achorutes muscorum* in Germany. The light was bright when the animals moved, less bright when at rest. If strongly stimulated an actual luminous secretion was formed. The light of these two springtails disappeared on death and was not due to luminous bacteria.

Flies. One of the few luminous animals sometimes living in caves is the famed New Zealand glowworm, *Bolitophila luminosa*, larva of a mycetophilid fly, first described by Hudson (1886). These larvae spin a glutinous web to which they cling. According to Hudson (1886, 1926), they do not always luminesce but "when disturbed they nearly always gleam very brilliantly for a few seconds, suddenly shutting off the light and retreating into the earth." The adult is also luminous on the tip of the abdomen. Wheeler and William (1915) saw a bluish green intracellular light from four parallel rods at the tip of the abdomen which proved to be the ends of the Malphigian tubules, the excretory organs of insects (Fig. 46). It is possible that these organs contain luminous bacteria. We are reminded of the many described cases of adult midges (Chironomidae) undoubtedly infected with luminous bacteria. They were first described by Hablizl (1782) and have been observed by Alenitzin (1875), Schmidt (1894), Henneberg (1899), Issatchencko (1911) and Behning (1929). If Bolitophila contains bacteria, they must be symbiotic, for every individual is luminous and the ends of the Malphigian tubules are enlarged into definite luminous organs. Wahlberg (1849) and Stammer (1932) described the larvae of luminous fungus gnats (a mycetophilid, Ceroplatus testaceus), living on glutinous webs on the under side of fungi (Fig. 44). The whole body was luminous in larval and pupal stages but the older adults were not. According to Stammer the light was not due to bacteria.

Moths. Isaac (1916) and Hepp (1927) both report that when a European moth, *Arctia caja*, is disturbed, a light organ is exposed from which exudes a drop of luminous pungent secretion. Light was also seen by Hepp in *Parascinia plantaginis* and in larvae of *Eudia pavonia* and *Zygaena ephialtes* var. *peucedani*, the exudation coming from warts or tubercles. Nothing is known of the structure of this luminous gland. Caterpillars infected with luminous bacteria have been referred to on page 29.

Beetles. Poets have missed the marvellous displays of many luminous creatures but great men—Shakespeare, Tennyson, Wordsworth, Lowell, Coleridge—and lesser ones have noticed the firefly and the glowworm.

"When evening closes Nature's eye,
 The glowworm lights her little spark

LIVING LIGHT

> To captivate her favorite fly
> And tempt the rover through the dark."[5]

> "Many a night I saw the Pleiads rising through
>   the mellow shade
> Glitter like a swarm of fireflies
>   tangled in a silver braid."[6]

Two groups of beetles possess intracellular luminescences, the elaterid or click beetles and the lampyrids, fireflies or lightning bugs. Glowworms may be either larvae of beetles or wingless female beetles. The larva may be luminous and the adult not, or vice versa. Some glowworms are purely aquatic, spending their life in pools and streams, breathing by means of tracheal gills oxygen dissolved in the water (Annandale, 1900; Blair, 1927; Okada, 1928). However, most glowworms are terrestrial. Reports that the carabid beetle, *Physodera noctiluca*, and the bupestrid, *Bupestris ocellata*, are luminous have not been confirmed. Neither has the report of luminous antennae in the beetle, *Paussus sphaerocerus*, from Sierra Leone. General accounts of insect luminosity are given by Seaman (1891) and Maluf (1938).

Every visitor to the West Indies has seen the large brilliant "Cucujo" beetle (Fig. 48) mentioned by Spanish writers as early as 1526. Bacon (1620), Bartolin (1647), Stubbes (1668) and Sloane (1707) also described it. Its history is given by Dubois (1886). Careful studies have been made by Kölliker (1859), Robin (1873), Heinemann (1872, 1873, 1886), Dubois (1886), Fuchs (1891) and many others. Dubois raised young larvae from the egg, which is itself luminous, and noted the tiny light organ near the head (Fig. 48).

The adult animal possesses two brilliant eye-like greenish luminescent organs on the prothorax and a ventral more orange

---

[5] Montgomery, *The Glow-worm*.   [6] Tennyson, *Locksley Hall*.

LIGHT-PRODUCING ORGANISMS

luminescent organ, on the first abdominal segment, visible only when flying. Sections of the organ show a mass of photogenic cells with granules immediately under the clear chitin, and below these, similar cells with fine granules that probably act as a reflector layer. When flying, this insect looks like a shooting star and Pickering (1916) estimated that its light at 53 meters (0.004 candle) was equivalent to a first magnitude star. Early accounts say that the natives of Spanish West India used them in place of candles and also kept them in their rooms to catch "gnattes." On festival days in June the cucujos were collected in great numbers and tied on the garments of young people who galloped through the streets at night. On such occasions the lover displayed his gallantry by decking his mistress with the living gems. *"Many wanton wilde fellows rub their faces with the flesh of a killed Cucuius with purpose to meet their neighbours with a flaming countenance and derive amusement from their fright."*[6a] Native girls used the insect for decoration in the hair or tied them to their feet to illuminate forest paths at night.

A rare beetle larva from South America, sometimes two inches long, with a red light like a glowing coal on the head, and rows of greenish lights on the sides of the body has been described by Azara in 1809 and from time to time since then (Reinhardt, 1854; Murray, 1868; Weyenbert, 1874; Burmeister, 1877; von Jhering, 1887; Haase, 1888; Barber, 1908). It is often called the railway bug. Such forms were originally named *Astraptor illuminator* by Murray and were by some considered larvae of *Pyrophorus*. They live in wood and are very active, "increasing and decreasing the light at will." Haas described a copulating male and female which definitely placed the animal as a female of the genus, *Phengodes*, related to fireflies. The males are winged with long branching antennae and

[6a] H. Sloane, "A Voyage to Jamaica," 1707, 2, p. 206-7.

not luminous. North American forms[7] lack the red light, and the larva and adult female are so nearly alike they can hardly be distinguished. The author has seen two live specimens, both of which glowed continuously night and day for several weeks. Fig. 49 shows one of these photographed by day and also by its own light. Note the row of lights on each side and a band of light across each segment.

In the true fireflies, the organs are ventral, on the last segments of the abdomen (Fig. 50). Many genera of fireflies exist and hundreds of species, mostly tropical; but in temperate North America, superb displays are visible from late May to early July. In Japan, fireflies are raised by dealers to be sold to young and old. A firefly festival is held on Lake Biwa near Kyoto. At a signal the merrymakers, who have rowed out in boats, free their pets from cages to fly high above the water and the shore. Again, a newspaper may report that ten thousand fireflies were liberated before the palace of the Emperor by school children as a token of affection.

The structure of the photogenic organ of the firefly has been widely studied (Kolliker, 1858; Schultze, 1865; Owsjannikow, 1868; Wielowiejski, 1882; Emery, 1884; Bongardt, 1903; Townsend, 1904; Vogel, 1915, 1921, 1922; Lund, 1911; Geipel, 1915; Williams, 1916; Dahlgren, 1917; Hess, 1922; Okada, 1935). It is made up of two kinds of cells, a dorsal mass of small cells several layers deep, the reflector layer, and a ventral mass of large cells with indistinct boundaries, the photogenic layer (Fig. 47). The photogenic cells, which contain granules, are divided into groups by large tracheal trunks which pass into the light organ and branch to form tracheoles connected with tracheal end cells. The exact distribution of tracheae varies in different species; but in all, the arrangement is

[7] These were originally described as the larvae of an elaterid beetle, *Melomactes piceus*.

such as to give a very abundant oxygen supply. The tracheoles pass into, and either end openly within the photogenic cells or anastomose with tracheoles from neighboring tracheae. Nerves, but no blood-vessels—for such are absent in insects— enter the organ. It is difficult to determine if the nerves supply the tracheal end cells or the photogenic cells.

The dorsal reflecting layer is made up of cells containing numerous minute crystals of some purin base, either xanthin or urates, which have a white milky appearance. While they are certainly not good reflectors in the optical sense, they do act as a white background, scatter incident light, and partially prevent its penetration to the internal organs of the firefly. In the larva the two layers are distinct and permanent from an early stage in development. They both come from the fat body and are mesodermal in origin, as first shown by Vogel (1913) and confirmed by many later workers. Only the tracheal elements are ectodermal. The fat body cells in the terminal end segments of the embryo may proliferate to form the organ or may wander into position from other regions. (Fig. 45)

Even the egg of the firefly (Lampyris) is luminous and glows with a steady light, and during the pupal period a definite light may be seen coming from all parts of the body (Spleist, 1647; Kratzenstein, 1757; Guéneau de Montbelliard, 1782). Some species take two years to develop. Curiously enough, the light organs of the larva of the firefly, two small dots of light on last segment of abdomen, are quite distinct from that of the adult (Fig. 44). Like so many other structures in insects, the adult organ is developed entirely anew from potential photogenic cells during the pupal period.

If the two small luminous dots of the larva of *Photuris pennsylvanica* (Fig. 44) are completely removed (Harvey and Hall, 1929) the animal shows no luminescence whatever until the latter part of the pupal period when diffuse luminescence, exactly like that of an unoperated pupa appears. The adult organ of such operated fireflies is luminous and perfect in every way. If the firefly light were bacterial in origin, removal of the organ should completely remove the bacteria and the adult should not be luminous. Since the adult is luminous, the granules in the organ of the firefly may be considered photogenic and not bacteria. The only alternate explanation is that supposed luminous bacteria might exist in a non-luminous form in some other part of the larva, but this does not seem probable.

Synchronous flashing of fireflies has excited as much interest and personal observation as the phosphorescence of the sea. There is no doubt of its occurrence, mostly in tropical countries, Siam, Burma, Malaysia, Philippines, East Indies, with a few reports from North America. Buck (1938) has prepared a comprehensive review of synchronous flashing with a full literature list. One of the earliest observations is by Kaempfer (1727) who described a display on the Meinan River near Bangkok, Siam. *"The Glowworms (Cicindelae) represent another shew, which settle on some Trees, like a fiery cloud, with this surprising circumstance, that a whole swarm of these Insects, having taken possession of one Tree, and spread themselves over its branches, sometimes hide their Light all at once, and a moment after make it appear again with the utmost regularity and exactness, as if they were in perpetual Systole and Diastole."*

Again Sir John Bowering (1857) remarks of the Siamese firefly: "They have their favorite trees, round which they sport in countless multitudes, and produce a magnificent and living illumination: their light blazes and is extinguished by a common sympathy. At one moment every leaf and branch appears decorated with diamond-like fire; and soon there is darkness, to be again succeeded by flashes from innumerable lamps which whirl about in rapid agitation."

These tropical displays are one of the outstanding sights of Siam and may extend for a quarter mile on trees along the bank of a river. Only a certain kind of tree (Sonneratia acida) is selected for the synchronous display, and its trunk may stand in several feet of water. The fireflies come out of the forest at dusk, flashing at random, and fly toward the trees. They flash at a rate of 100 to 120 a minute and the display may continue "hour after hour, night after night for weeks or even months" without regard to wind or weather. Only in bright moonlight does the display cease (Smith, 1935), and Morrison (1929) found that playing a flashlight over the insects would stop the rhythm. Most observers have not noticed a pacemaker, but Alexander (1935) mentions a leader and a wave of light which proceeds very rapidly over the colony.

If animals tended to flash in a definite rhythm, which they do, and also if each immediately flashed when stimulated by a flash from another at a time near the natural period of the rhythm, all members would be soon brought to synchronism. Pacemakers might appear here and there (Richmond, 1930). The mechanism would be similar to rhythmic discharge of nerve cells and quite understandable since the flash is nerve controlled in the firefly. However, the reason for rhythmic flashing in the tropical fireflies is still uncertain, since both Smith and Morrison report that all the insects flashing in unison are males. The females presumably lie hidden in the jungle and the light seems to have nothing to do with mating.

In American and European fireflies, synchronism is rare but can occur, due to the fact that the light is a mating signal. The use for mating seems to have been early recognized and was carefully studied by Osten-Sacken (1861), McDermott (1910-1917), Mast (1912), Hess (1920) and Buck (1935, 1937).

Most American fireflies keep hidden under leaves during the day but come out at dusk, when the illumination reaches a certain low value. The females (sometimes wingless) usually remain in the grass and the males fly about. The males flash and the females respond. Then the males fly toward the females and again flash when the females again respond. In 5 to 10 flashes the pairs have found each other. With great truth Bishop Heber in his *Tour through Ceylon* remarks:

"Before beside us and above
The firefly lights his torch of love."

Each species of firefly has a characteristic method of flashing, chiefly distinguished by intensity, duration, number and interval between flashes. The color may also be different. In *Photinus pyralis* the essential feature of the flashing signals is that the responding female invariably flashes two seconds after each successive rhythmic flash of the male (Buck 1935, 1937). Several males are attracted to the same female and flash in unison. This little group acts as a unit in stimulating another female some distance away. The second female then attracts males and a second unit becomes synchronous. The process can be initiated by substituting a flashlight for the female, a method much used in studying the mating of fireflies. Buck (1937) has shown that males will respond to other males or to a flash of light of varying intensity or duration and of any color from green to deep red (not blue or blue green), provided the time interval is two seconds after a previous flash.

MOLLUSCA

This large group contains only three certain luminous species[8] if we except the squid, which only an expert would recog-

---

[8] A mollusc, *Triopa (Euphurus) fulgurans*, from New Caledonia is said to produce short bright flashes of light (Risbec, J., C.R.Ac.Sc., 181, 472, 1925).

nize as a mollusc. These are the boring clam, *Pholas dactylus*, the pelagic nudibranch, *Phyllirrhoë bucephala* and another nudibranch, *Plocamopherus ocellatus* (Eliot, 1908). Among deep-sea squid there are many luminous species with complicated lantern-like luminous organs. Giglioni (1870) says the pteropods, *Cleodora* (= *Clio*) *cuspidata*, *Hyalea* (= *Cavolina*), *Styliola* and *Creseis conica* luminesce and Champion (1883) records a *Helix* but the validity of these reports is very uncertain.

Pholas, the rock clam, may be compared to the worm, *Chaetopterus*, as a luminous form lying hidden away in mud or soft rock, extending only the siphon to obtain its stream of fresh sea water. Common in the Mediterranean, it has been known since the earliest times (Pliny) and was studied by Réaumur (1723) and Beccari, Monti and Galeati (1724). Many of their experiments were directed to preserving the luminous material and they recommended making a paste with flour or adding honey to the luminous parts. Material preserved in this way would again give light even after a year when poured into water. Panceri (1872) and Dubois (1887-1917) have carried out extensive chemical studies and *Pholas* played a large part in Dubois's theory of bioluminescence as an enzymic (luciferase) oxidation of a definite compound, luciferin.

Five regions of the animal are luminous, as shown in Fig. 51. The histology has been studied by Dubois (1894), Forster (1914) and Dahlgren (1916). A slime is produced from large gland cells with long ducts. The light comes from granules, which dissolve with luminescence, and may be bright enough to illuminate a jar of sea water. Nerves go to the organs and the luminous secretion is shot out of the siphon, the only way in which the light could be seen by other creatures without tearing open the animal.

Phyllirrhoë, the flowing leaf, a pelagic transparent shell-less snail found in the Mediterranean Sea and Atlantic Ocean, has always excited the interest of zoologists (Fig. 52). It was first observed to luminesce by Panceri (1873) and has been studied by Vessichelli (1906), Born (1910) and especially by Trojan (1910). The animal lights only on stimulation, most brightly on the head end and contours of the body. The light comes from small points that flash and die out. These represent gland cells, single or in groups, with openings to the exterior, but an actual secretion has not been described. A nerve goes to the cell and the light is readily obtained by electrical stimulation.

Cephalopods. Like the coelenterates, squid are a class of fundamentally luminous animals. They have developed the most varied methods of lighting, from the harboring of luminous bacteria in *Sepiola*, *Rondoletia*, *Loligo* and *Euprymna*, previously described, to the most complicated lanterns which culminate in the deep sea, *Lycoteuthis diadema*, described by Chun (1903, 1910) or *Nematolampas regalis* of Berry (1913). According to Chun, Verany (1851) was the first to observe a luminous squid, *Histioteuthis bonelliana*, in 1834 but our real knowledge of deep-sea fauna came with the voyage of the *Challenger*. Joubin (1893-1912), Hoyle (1902-1909), Chun (1900-1913), Berry (1913, 1920, 1926), Mortara (1917-1924) and Naef (1912, 1921) have dealt with luminescence in this group. Berry (1920) has listed all known luminous species among the cephalopods and collected the facts concerning light emission. Only two octopods are luminous, *Melanoteuthis luceus* and *Eledonella alberti*. The decapods are divided into the oegopsid and the myopsid squid. In the former, 39 genera out of 60 are luminous; among myopsids only 6 out of 27. It is in this latter group that the luminous organ is open and that luminous bacteria are associated with light production. A possible exception is the closely allied but extraordinary deep-sea form, Heteroteuthis dispar, first studied by Meyer (1906, 1908).

Heteroteuthis is a small myopsid squid living in a depth of 1,200 to 1,500 meters in the Mediterranean. Its luminous organ is large and unpaired lying between the gills and anterior to a small ink sac. The author has seen this animal at Messina, Sicily. When disturbed, it surrounds itself with a luminous secretion instead of ink, a spectacular display. The glowing secretion is shot out through the siphon as a cloud of luminescence. We can picture attacking fish subjected to a veritable bombardment of liquid fire, quite as startling if not as dangerous as any developed during the World War. Squid ordinarily surround themselves with "ink," one of the blackest of fluids. It is paradoxical indeed to find a squid manufacturing not only a clear fluid but one actually shining with its own light.

The gland is surrounded by rather thin epithelial cells that do not look secretory in character, and blood vessels pass in and out of the interior, which is filled with a mass of rods, 3 μ in diameter and 2-7 μ long, embedded in a cementing ground substance. These appear to divide or bud and were assumed to be luminous bacteria by Pierantoni (1918, 1924, 1926). Mortara (1922, 1924) and Skowron (1926) believe otherwise. If these granules are bacteria, they are so modified that they should form an entirely separate group. Their luminescence in sea water does not persist over three hours, while luminous bacteria from Sepiola retain their luminescence over twenty-four hours. Myopsid squid are shown in Fig. 55.

In oegopsid squid the light is intracellular and the organs have no ducts or openings and are often of many different types (Fig. 58). Usually they are ventral in distribution. The arms and especially the eyeball are favorite positions. In fish also, a subocular luminous organ is common and we may raise the question without offering a logical answer, as to why there

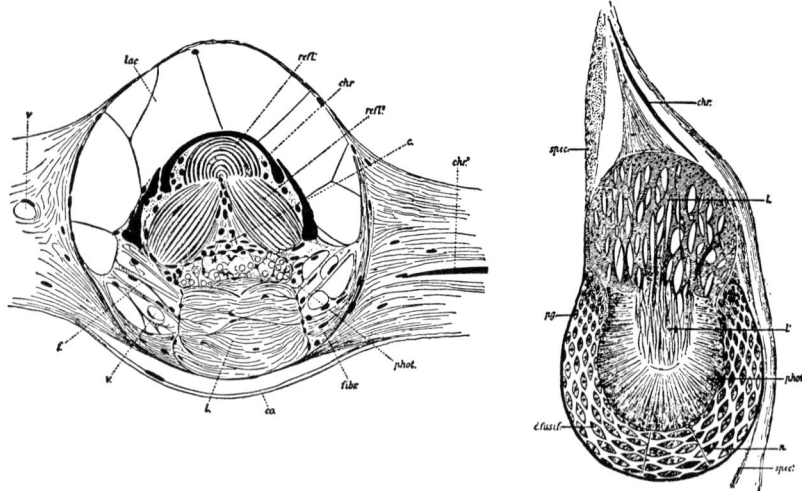

FIG. 58 Transverse sections of the light organs of two squid, *Abraliopsis* (left) and *Calliteuthis* (right). After Chun.

is so frequently a close association of light-emitting and light-detecting organs.

Most oegopsid squid are only occasionally obtained, in deep-sea collecting (Fig. 53), but one of them, *Watasenia*, has been extensively studied in the living state. This is the firefly squid, "hotaru-ika," *Watasenia scintillans* (Fig. 57), caught in immense numbers (1,000 tons) in the spring and early summer in Toyama Bay on the west coast of Japan. They are used to fertilize the rice fields. The first observations of this squid were made by Watase (1907) followed by those of Ishikawa (1913), Sasaki (1914), Shoji (1919), Kishitani (1928), Tagaki (1933) and Okada, Takagi and Suzimo (1933). The author observed (1916) great net loads of these squid waving their arms and flashing the terminal lights in a brilliant, decidedly bluish luminescent display. Of the three kinds of light organs shown in Fig. 57 the arm organs are brightest, the five on the lower side of the eye ball next, and the minute dots scattered over the ventral surface of the body dimmest. The sexes differ in number of minute scattered organs. Since the migration of this squid to the surface in the spring is for breeding, we may presume their use is for sex attraction.

Chun (1910) had the good fortune to observe a deep-sea form, *Lycoteuthis diadema*, caught from the *Valdivia* in the Indian Ocean at 3,000 meters depth. It was kept a sufficient time in iced sea water to see that light from the central eye organs was ultramarine blue, that from the lateral ones pearly white, while the median ventral organ shone sky blue and the anterior organs ruby red, an extraordinary spectacle. It is not definitely known whether some of the various colors are due to chromatophores acting as absorption screens. This form is shown in Fig. 54. Probably the most elaborate arrangement of light organs is found in *Nematolampas regalis* described by Berry (1913) from the Kermadec Islands. It has over 90 good-sized organs, most of them arranged regularly on the extra long ventrolateral arms. No one has ever seen the light of this squid.

ECHINODERMATA

This phylum, in which luminescence might hardly be expected, is made up of five classes—sea lilies, starfish, brittle or snake stars, sea urchins, and sea cucumbers. Only the brittle stars are luminous. Viviani (1805) appears to have been the first to see their light, but many followed him and the process has been carefully studied by Mangold (1907), Sterzinger (1907), Trojan (1908), Reichenspergers (1908) and Sokolow (1909). Trojan noted that the young plutei were luminous.

The luminous genera are *Amphiura*, *Ophiopsila*, *Ophioscolex*, *Ophiacantha*, *Ophiothrix* and *Ophionereis*. The root *ophio*, a snake, supplies the name "snake stars" from the snake-like movements of the arms (Fig. 56). These forms live under rocks or among seaweeds. When disturbed, the arms and body light, the exact location depending on the species. It may be emitted from plates, spines or even tube feet, according to Trojan and Mangold. No secretion is discharged that can be rubbed off on the fingers but the photogenic gland cells are described as having long ducts which appear to open on the outside.

CHORDATA

Although this phylum contains the true vertebrates with definite backbones, some of the lower members, the balanoglossids and the tunicates, would never be recognized as related forms.

Balanoglossids. These creatures are perhaps ancestors of vertebrates. They are soft worm-like animals, living in sand or under rocks. They glow with a weak diffuse light over the body when disturbed. Less than a hundred species are known. The genera *Balanoglossus* and *Ptychodera* are luminous but *Dolichoglossus kowalewski*, occurring at Woods Hole, Mass., is not. Panceri (1875) appears to have been the first to observe the light of a Naples form, *Balanoglossus minutus*. The whole body secretes a luminous slime from unicellular mucous and photogenic cells, which comes off on the fingers.

Tunicates or *Ascidians.* Reports of luminescence in *Salpa*, *Doliolum* and *Appendicularia* have been made from time to time. *Salpa* and *Doliolum* might be expected to luminesce although the author has never seen a luminous individual. The luminescence of *Appendicularia* may be considered doubtful at present. Stier (1938) has studied the embryology of *Salpa* in fixed material and believes that symbiotic luminous bacteria infect certain blastomeres which then disintegrate. The liberated bacteria enter follicle cells which have migrated into the embryo and which later form the luminous organ. The sedentary tunicates, *Ciona* and *Botryllus*, have also been reported luminous but the accuracy of these observations is very doubtful.

A near relative, *Pyrosoma gigantium*, the "fire cylinder," is a well known luminescent form, early described by Peron (1804) and Lesueur (1815) and observed by many. Panceri (1872-1873) made a careful study.

The animal, a colony of small tunicates, may reach a length of several feet and gives a brilliant intracellular display of light. Each individual possesses two masses of luminous cells at its "shoulder," well seen in Fig. 59. The cells are rounded and filled with bacteria-like inclusions, sometimes branching, studied by Julin (1909, 1912), and regarded as symbiotic bacteria by Pierantoni (1921, 1923) and Buchner (1914, 1930).

According to Panceri and also Polimanti (1911) the animal luminesces on stimulation but not if recently dead. Like many other forms, luminescence occurs if moistened after drying. In fresh water a colony lights continuously for some time. The luminescence can be conducted from one individual of a colony to another, just as in pennatulids. There is possibly nerve communication between the individuals, although they also respond to light stimulation. The luminescence of one individual can set off the next so that a wave of luminescence travels over the colony by successive light stimulation of adjoining individuals. Burghause (1914) has demonstrated that this can actually occur. A colony in one glass dish (A) will cause a second colony (B) in another glass dish to light. The luminescence first appears on the animals of B toward the A colony and then spreads over other animals of B. The promptness of response depends on how near the colonies are. Daylight inhibits the luminous response. The light from other luminous animals, or even a lighted match, can set off the luminescence of *Pyrosoma*. A pigmented organ near the ganglion must act as an eye to stimulate nerves which relay the impulse to the light organ. Nerves have not been demonstrated histologically, but must be present. The use of the light is problematical, as a lighting *Pyrosoma* is said to be readily eaten by fish and crabs.

Fish. Curiously enough fish were not generally recognized as possessing the power to emit light until relatively recently, although a number of surface fish are luminescent. Deep-sea fish were not available and Science of the 1860's never dreamt of the wealth of bizarre forms soon to be discovered (Fig. 62).

The first systematic attempts to study life at any depth were made by Edward Forbes, who in 1841 dredged and found material at 230 fathoms but who believed no life existed below 300 fathoms. Michael Sars in Norway and C. Wyville Thomson in England continued this work. Thomson in the *Lightning*, attained 600 fathoms in 1868, and in the *Porcupine* in 1870, reached 2,435 fathoms, even there to find abundant living material. The *Challenger* expedition from 1872 to 1876 opened up a new world of remarkable deep-sea creatures, many luminous forms among them. Fishes were obtained at 2,750 fathoms, although life is more abundant at 400 fathoms and in the twilight zone. The human eye can distinguish nothing below 300 fathoms (Beebe, 1934). The greatest depth yet measured is about 5,800 fathoms.

Two groups of fishes light, the cartilaginous selachians and the bony teleosts. One electric ray, *Benthobatis moresbyi*, dredged from 430 fathoms off the Indian coast is reported by Alcock (1902) to have no functional eyes and a row of luminous spots at the edge of the body. Risso (1810) noted that *Chimaera arctica*, an elasmobranch, poured a lighting slime from pores in its snout. Luminous sharks are better known and were first described by Bennett (1840) who observed the light from an Australian species, kept in an aquarium, while making a whaling voyage around the world.

In the Straits of Messina, deep-sea fish, teleosts, are swirled up from the depths and Cocco (1838) realized that the rows of spots on their sides (Fig. 65) were "punti luminosi," not accessory eyes as was later (Leuckart, 1864) supposed. Although Reinhardt (1854) saw a fish, *Astronesthes*, which emitted vivid greenish lights intermittently, during a voyage to Brazil, others, studying preserved material, thought the photophores were electric organs or organs of taste or of a sixth sense. During the 'seventies so many zoologists had seen the light that their function became definitely established. Brauer's (1906, 1908) monumental work on the structure of the organs eclipses all else. He recorded luminous organs from 239 species of fish belonging to 69 genera. The structure of the organs has also been studied in a large number of forms by Leydig (1879), Usow (1879), Emery (1884), Mangold (1907), Ledenfeld (1887, 1905), Greene (1899, 1924), Brandis (1899), Chiarini (1900), Handrick (1901), Gatti (1904), Steche (1909) and Trojan (1915). Sanzo (1912-1925) has studied the embryology of many luminous bony fish. Beebe (1937) has collected deep-sea fish off Bermuda representing 98 genera, about one-third of all known genera of deep-sea fish. Eighty-one per cent of these genera contained some luminous species.

Johann (1899), Burckhardt (1900), Leydig (1909), Oshima (1911) and Dahlgren (1917) have dealt with the Selachians. The genera *Isistius, Somniosus, Etmopterus* and *Centrosyllium* are known to contain luminous members.

In these fish the minute punctate photophores are densely arranged in patterns. Sometimes there may be 70 per square millimeter while in other regions only a few. Each photophore is pearly in the living condition, surrounded by black pigment. It lights only when the fish is handled and the skin rubbed or electrically stimulated. The luminescence appears slowly. This suggests that possibly the luminescence in this fish is due to a hormone secretion which in its turn stimulates the photophores (see p. 85). A much needed physiological investigation should be undertaken as the organs might harbor luminous bacteria whose light appears when pigment cells retract.

Structure of the light organs of fishes is well known but details of the method of lighting and physiology are not. We can recognize a number of groups. There are (1) fish such as *Photoblepharon, Anomalops, Monocentris, Equula*, and possibly *Ceratias*, already described (pp. 31-4), in which the light is due to symbiotic bacteria which remain in the organ. We can also recognize (2) fish such as *Malacocephalus, Coelorhynchus* and *Physiculus* (p. 34), described by Osorio (1912), Hickling (1925, 1926, 1931) and Haneda (1930), in which a gland with a duct on the ventral side of the body forms a viscous luminous secretion which smears the belly of the fish when handled. These fish, when disturbed, presumably give off a puff of light and swim in another direction. The secretion uses a great deal of oxygen and evidence has already been given (p. 34) to indicate that the granules of the secretion are luminous bacteria. They are surrounded by a membrane which behaves in some ways like that of a typical cell. The organ of *Acropoma* contains luminous bacteria and has a long duct, but the ejection of luminous bacteria has not been described. Then there are (3) true deep-sea forms with rows of photophores, lantern-fish, *Myctophum, Maurolicus, Chauliodus* (Fig. 60) and *Stomias*, the hatchet fish, *Argyropelecus* (Fig. 61), and *Astronesthes*, "eater of stars." In these, a cup-shaped mass of photogenic tissue is partly surrounded by a pigment layer and capped with a clear region, more or less lens-shaped, through which the light shines (Fig. 63). Ordinarily no duct is present. Frequently there is an organ just below the eye, and in those forms which have long tentacles or barbels, there is an organ at the tip of the tentacle or barbel. Finally, we find (4) deep-sea fish like *Cyclothone* (Fig. 64) and *Gonostoma*, with luminous organs similar to those in the third group but with a long duct leading to the exterior (Fig. 63). As some of these organs possess lenses (Fig. 65), the meaning of the duct is problematical.

Despite the formidable structural array of light organs, descriptions of the light are few and not very startling. Mention has already been made of the slow appearance of light in sharks. *Maurolicus* was observed by Mangold (1907) to light only on stimulation or when placed in fresh water. Oshima (1911), speaking of *Maurolicus*, says, "the yellowish green light was very feeble, not strong enough to illuminate any object held near by." *Myctophum* gave a momentary weak bluish luminescence "like an electric spark." Skowron (1928) saw the beautiful luminescence of *Chauliodus* at Messina, but often no light can be obtained from deep-sea fish, especially if they are in poor condition, as frequently happens after being brought up from great depths.

Greene (1899, 1924) has studied the California singing fish, *Porichthys notatus*, caught at the spawning season in tide water in Monterey Bay. No luminescence was observed in the free swimming fish, but on strong mechanical or electrical stimulation or addition of ammonia water, light will appear. At least 750 photophores are well developed and there may be as many as 840. The light is slow to appear (8-10 seconds) after electrical stimulation, although it lasts 20 seconds. This led Greene to try injection of adrenalin, which gave immediate luminescence of organs near the point of injection and finally spread to other regions until the whole fish was glowing brilliantly and remained glowing for over an hour. The same effect can be obtained at other times than the breeding season and from both males and females. Insulin was without effect but the experiments showed that light in this fish is definitely hormone controlled. We may compare its light production to color changes in other fish, also controlled by hormones.

The author has seen the light of *Photoblepharon*, *Anomalops*, *Monocentris* and *Echiostoma tanneri*. Only the first two were striking in appearance. The bacterial light on the lower jaw of *Monocentris* is so faint it is hardly noticeable. *Echiostoma* caught at 800 fathoms in one of Beebe's hauls south of Bermuda, is large for deep-sea fish, a foot long and black in color (Fig. 65). The specimens were brought to the laboratory in good condition in iced sea water. They have a prominent cheek organ, pinkish in color, which flashed with a decidedly bluish luminescence when the fish was handled, especially when lifted out of the water. No other luminescence of any kind could be noted, however, despite the fact that the fish was squeezed and twisted to stimulate it strongly. A hypodermic needle was then inserted but no luminescence additional to that of the cheek organ appeared. However, when a little adrenalin (1:1000 in physiological salt) was injected with the hypodermic into the side, about one-third toward the tail end, there immediately appeared a yellowish luminescence of photophores locally, near the point of injection, and soon practically all of the photophores of the fish were luminescing with a yellowish, moderately intense, continuous glow. This lasted a few minutes and then went out and could not be again excited by rubbing or handling, but appeared as before on a second, third and fourth injection of adrenalin. The last injection was large and excited all organs and also the pectoral and ventral fins. There is no doubt of the luminescence of these fins despite the fact that they do not possess any marked luminous organs. No luminescence was observed in the tail, anal fins, long pectoral rays, or barbel on lower jaw. The cheek organ flashed at intervals after adrenalin injection but did not change in rhythm or in any noticeable way. The flashing of this organ is not due to unscreening of a continuously luminous surface. The light appears and disappears in the organ itself and for this reason we may presume that *Echiostoma* is self-luminous and does not harbor luminous bacteria as is the case in the cheek organ of *Photoblepharon* and *Anomalops*.

It should be mentioned that adrenalin is not a stimulant for light production after a fish has been dead some time. Other dead deep-sea fish, and even a feebly moving *Linophryne arborifera* (Fig. 65), could not be made to light by injecting adrenalin (Harvey, 1931). Skowron (1928) also was unable to obtain luminescence by adrenalin injection into a *Chauliodus sloanii* which had just died.

Beebe's (1926) observations on *Myctophum coccoi* led him to believe that the lights are used to distinguish the species and the sexes, for illumination and possibly also to blind other fish; a sudden bright flash on the dark adapted eye leaves it momentarily blinded, when the fish might quickly make its getaway. Concerning the luminescence of such bizarre forms as are shown in Figs. 66 and 67, nothing is known.

Fish are the highest group of animals in which self-luminosity is known. All the reported cases of light from amphibia, reptiles, birds and mammals have been due to some secondary phenomenon.

CHAPTER III

TYPES OF LUMINESCENCE

FUNDAMENTALS

NEWTON thought of light in terms of corpuscles; Huygens in terms of waves; today corpuscles have the properties of waves. It is therefore convenient to picture the light which affects the eye as a very small segment of an enormous spectrum of waves with radio at one end and cosmic rays at the other (Fig. 68). The velocity is 300,000 kilometers a second, product of wave length (λ) and frequency (ν). Some phenomena of light can be explained only if the waves are emitted in units, quanta ($h\nu$), which represent a universal constant of energy, h,[1] multiplied by the frequency. The concept is useful in photochemical reactions where the primary effect of light may be thought of as due to one quantum causing change in one molecule. Conversely, in a chemiluminescence, one molecule undergoing change should emit one quantum. However, such high efficiency is never found.

It is generally accepted that light emission is connected with movement of the electrons making up atoms and molecules. These are the ultimate oscillators. The electron is displaced by various kinds of energy (electric, mechanical, chemical, thermal, etc.) to another (higher) energy level. Such an atom or molecule is said to be "excited." In returning to its original level, the excess energy appears as light. The life of an excited state is usually very short, measured in hundred-millionths of a second, although some "metastable" states last

---

[1] The value of $h = 6.547 \times 10^{-27}$ erg.-sec.

TYPES OF LUMINESCENCE

much longer. Displacement of an electron may be partial or complete, when ionization of the atom occurs. Since these changes easily occur, it is not surprising to find light produced by the many methods mentioned in Chapter I.

In an atom, which is relatively simple, only a few electron shifts are possible. Consequently atoms emit light at discrete frequencies. Line spectra are obtained whose complication depends on the structure of the atom. Iron atoms emit thousands of lines. Molecules emit groups of lines or bands, even continuous spectra, if the molecules are large and complicated. The spectra of all luminous organisms are continuous over a short range of wave lengths, meaning that molecules of some fairly complex substance are emitting the light. The emission occurs at low temperatures.

In solids at high temperatures, the energy for light emission comes not only from electron transitions but also from the kinetic energy of the moving molecules themselves, directly due to the temperature of the source. Many frequencies are emitted and a continuous spectrum over a very wide range is obtained, that of incandescence. Since all radiation is due to "excited" atoms or molecules, i.e. those which have an electron displaced or lost, bioluminescence must result from some one of the known means of shifting electrons.

Sir Isaac Newton was familiar with many ways of producing light. After speaking of incandescence as due to vibratory motions of bodies he questions in his *Opticks* (1704) as to whether other light emissions are not similar vibratory motions:

Query 8. *"Do not all fix'd Bodies, when heated beyond a certain degree, emit Light and shine; and is not this Emission perform'd by the vibrating motions of their parts? And do not all Bodies which abound with terrestrial parts, and especially with sulphureous ones, emit Light as often as those parts are* sufficiently agitated; whether that agitation be made by Heat, or by Friction, or Percussion, or Putrefaction, or by any vital Motion, or any other Cause? As for instance; Sea-water in a raging Storm; Quick-silver agitated in vacuo; the Back of a Cat, or Neck of a Horse, obliquely struck or rubbed in a dark place; Wood, Flesh and Fish while they putrefy; Vapours arising from putrefy'd Waters, usually call'd Ignes Fatui; Stacks of moist Hay or Corn growing hot by fermentation; Glow-worms and the Eyes of some Animals by vital Motions; the vulgar Phosphorus agitated by the attrition of any Body, or by the Acid Particles of the Air; Amber and some Diamonds by striking, pressing or rubbing them; Scrapings of Steel struck off with a Flint; Iron hammer'd very nimbly till it becomes so hot as to kindle. Sulphur thrown upon it; the Axletrees of Chariots taking fire by the rapid rotation of the Wheels; and some Liquors mix'd with one another whose Particles come together with an Impetus, as Oil of Vitriol distilled from its weight of Nitre, and then mix'd with twice its weight of Oil of Anniseeds. . . ."

To Newton, "shining fish" and "vulgar phosphorus" were of as much interest as light from heat or any other cause. The surest way to obtain a light is to increase the temperature, when temperature radiation or incandescence results.

INCANDESCENCE

Every solid heated to about 525°C. gives off rays that are just visible as a faint red glow to the eye, of wave length 0.76 micron.[2] As the temperature increases still shorter wave lengths become apparent, and the light changes to dark red, cherry red, dark yellow, bright yellow, white-hot and blue white. We

---

[2] Since wave length (λ) x frequency (ν) = velocity of light, 300,000 kilometers per second, various waves can be designated either in terms of wave length or frequency. Throughout this book wave lengths expressed in fractions of a micron, 0.001 millimeter (25 mm. = 1 inch), will be used.

TYPES OF LUMINESCENCE

can speak of a color temperature. Shorter and shorter wave lengths appear until those below 0.4 micron again fail to affect our eye. They are very active in producing chemical changes but we have no sense organs for perceiving them. Thus, a white-hot object liberates radiant energy or radiant flux of many different wave lengths, corresponding to what used to be called "heat, light and actinic rays." All can be dispersed by prisms of one or another appropriate material to form a wide continuous spectrum, such as that indicated in the center of Fig. 68, including infrared, visible and ultraviolet rays.

So closely associated are light and heat in all our experience that the two appear inseparable. "Hot light" seems tautological, "cold light" is paradoxical. Most illuminants in use today are patterned after the sun and stars. The attempt is made to heat a solid filament to the highest temperature possible. This is an application of the first law of incandescence (Stefan-Boltzman Law), that the total radiant energy emitted increases with the temperature and, for a perfect radiator or black body, varies as the fourth power of its absolute temperature, T.

A second law (Planck's Law) relates the radiant energy to the wave length. More energy is emitted at one particular wave length (λ_{max}) than at longer or shorter ones, also depending on the temperature. If the various waves are intercepted in some way, their relative energy can be measured by an appropriate instrument, and spectral energy curves can be drawn, showing the distribution of energy throughout the spectrum. Fig. 70 gives a few of the curves, and it will be noted that the maximum shifts further toward the shorter waves the higher the temperature. In fact, for a black body, λ_{max} x T = 2890 (Wien displacement Law), and at 6000°C. (about the surface temperature of the sun) λ_{max} lies within the visible spectrum. In gas or electric lights it lies in the

LIVING LIGHT

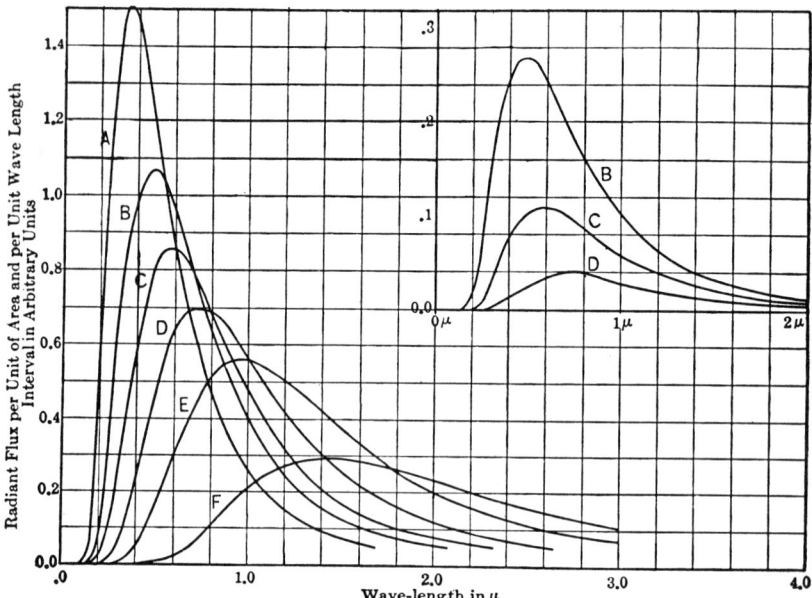

Fig. 70 Spectral distribution of black-body radiation at different temperatures, to illustrate the radiation laws. A, 8000°K; B, 6000°K; C, 5000°K; D, 4000°K; E, 3000°K; F, 2000°K. Wave lengths along horizontal; energy along vertical. In upper right the energy values are in proper units. Note that relatively little radiation is emitted in the visible region at the lower temperatures; also the shift in maximum of emission with increase in temperature. Reprinted by permission from Cady and Dates, *Illuminating Engineering*, published by John Wiley and Sons, Inc.

infrared region. The area enclosed by these spectral energy curves represents the total energy emitted, and, knowing this and the area enclosed by the curve for visible radiation, it is easy to determine how efficient a source of light is as a light-producing body. We shall inquire more fully into this question in Chapter VI in considering the efficiency of the firefly as a source of light.

The light coming through a small hole in an incandescent sphere is perfect black body or temperature radiation; that

TYPES OF LUMINESCENCE

from the surface of a carbon filament is nearly perfect, while that from a tungsten filament does not quite conform to the laws of radiation. Tungsten is called a grey body and emits less light than a black body of the same temperature but no one could mistake a white-hot tungsten filament for any other light than an incandescence. Its molecules are excited by the high temperature and tungsten gives a continuous spectrum. A candle flame is also an incandescence due to the temperature of solid carbon particles, but all flames are not incandescent and some may have very low temperatures.

Newton brings this out in two queries concerning the relation between the light of flame and of solids. He asks:

Query 9. "Is not Fire a Body heated so hot as to emit Light copiously? For what else is a red hot Iron than Fire? And what else is a burning Coal than red hot Wood?

Query 10. "Is not a Flame a Vapour, Fume or Exhalation heated red hot, that is, so hot as to shine? For Bodies do not flame without emitting a copious Fume, and this Fume burns in the Flame. The Ignis Fatuus is a Vapour shining without heat, and is there not the same difference between this Vapour and Flame, as between rotten Wood shining without heat and burning Coals of Fire?"

Newton's selection of *Ignis Fatuus* may be questionable but the difference between a vapor shining without heat and a candle flame is perfectly valid. Many years later it was suggested by Wiedemann (1888, 1889) that any light not due to temperature radiation be called a luminescence, in contrast with incandescence, and that suggestion is universally followed today. A literature has appeared in the field of luminescence that is completely overwhelming. It reached its height when the monographs of Pringsheim (1928), Lenard, Schmitt and Tomaschek (1928), Nichols, Howes and Wilber (1928), Tomaschek (1928) and Gudden (1929) appeared. The older publications of Nichols and Merritt (1912), Baly (1915) and Merritt, Nichols and Child (1923) should be consulted as well as the encyclopaedic work of Kayser (1908) in which the history of the subject is recorded fully. Two recent symposia on luminescence, one held in Warsaw in 1936 and one in London in 1938, present the latest viewpoints. The latter, together with the report of a Committee of the National Research Council on Chemiluminescence (1927) containing papers by Harvey, Adams, Garrison, Pfund and Taylor, will serve for reference to work in this field.

LUMINESCENCE

Although perhaps illogical, it is convenient to classify cold lights according to the method of producing them. The following kinds of luminescence have been described:

    Candoluminescence
    Pyroluminescence
    Thermoluminescence
    Phosphorescence and Fluorescence
        Photoluminescence
        Cathodoluminescence
        Andoluminescence
        Radioluminescence
    Electroluminescence
    Sonoluminescence
    Galvanoluminescence
    Triboluminescence or Piezoluminescence
    Crystalloluminescence and Lyoluminescence
    Chemiluminescence or Oxyluminescence
    Bioluminescence or Organoluminescence

These various luminescences are so interrelated that some could be quite logically described under several different

TYPES OF LUMINESCENCE

heads. Thus we have such expressions as electro-photoluminescence, sonic-chemiluminescence and tribo-thermoluminescence. The reason is easy to see. If a compound contains molecules with easily displaced electrons it may be excited to luminesce in a number of different ways. Indeed, a compound which luminesces as a result of exposure to light, we may predict will be susceptible to other modes of excitation. This is sometimes expressed by saying that it contains luminophore groups. It is evident that luminous animals have been able to select and synthesize a substance with easily excited molecules.

Candoluminescence. When some bodies, such as zinc oxide, are heated, their temperature becomes higher and they give off shorter wave lengths than would be expected. The distribution does not conform to the radiation laws. Nichols and Snow (1892) referred to this excess of short wave lengths, whose crest is often in the violet region, as a luminescence, and it later came to be known as candoluminescence, the luminescence of incandescent solids. The extraordinary actinic activity of the magnesium flame, or of rare earth oxides used in the Welsbach mantle, are examples. Nichols, Howes and Wilber (1928) regard the effects as fundamentally due to fluorescence excited by the incandescent radiation.

Pyroluminescence. Another example of luminescence at fairly high temperatures is flame luminescence or pyroluminescence, the various colors which appear when salts are held in a bunsen burner, yellow of sodium, red of lithium, green or blue of copper. This light accompanies excitation of atoms or of molecules; for example, recombination of ions to form molecules, fundamentally a chemical reaction. In general the light of flames may be due (1) to the pyroluminescence just referred to (2) to reflection from solid particles in the flame or (3) to incandescence of carbon or other solid particles.

LIVING LIGHT

Some very low temperature flames are known, such as that of carbon disulphide in air, rich in ultraviolet rays, despite its relatively low temperature. While these flames are of interest to the physicist and chemist (see p. 113), they can have no direct bearing on the luminescence of animals, and their consideration will also be omitted. (See Bancroft and Weiser, 1914-1915.)

Thermoluminescence. Robert Boyle (1662) first noted that diamonds gave off light when heated slightly. This is thermoluminescence. Many minerals, marble, apatite, quartz and especially fluorspar are examples. This last mineral was known to Elsholtz (1681) as *phosphorus smaragdinus*, and both Oldenburg (1705)[3] and Homberg (1694) noticed its thermoluminescence. Only certain varieties of fluorspar such as chlorophane show the phenomenon well. A crystal of one of these varieties heated in the bunsen flame on an iron spoon will give off a white light long before any trace of redness appears in the iron. Other crystals may luminesce in hot water.

In every case this thermoluminescence is dependent on some previous illumination or radiation of the crystal. If kept in the dark for a long time no trace of light appears when chlorophane is placed in hot water, but after a short exposure to the light of an incandescent lamp, although no light can be observed in the fluorite at room temperature, quite a bright glow appears at 100° C. Calcium, barium, strontium, magnesium and other sulphates containing traces of manganese, show a similar phenomenon after exposure to cathode rays. They emit light during bombardment, but this soon ceases when the rays are cut off. If the sulphates are now heated they give off light and this power may be retained for months after the exposure to cathode rays. One of the most striking thermoluminescent effects can be obtained by exposing fluorescent

[3] The actual observation was earlier, as Oldenburg died in 1678.

TYPES OF LUMINESCENCE

screens to cathode rays, gamma rays or X-rays at liquid air temperatures. No light may appear at this temperature but as the screen warms, luminescence in a series of colors occurs. Barium platino-cyanide screens exposed to X-rays at liquid air temperatures and then warmed, passed through six luminescent periods emitting successively light of green, yellowish green, reddish brown, light green, light reddish brown, and dark reddish brown color (Coolidge, 1925; Wick, 1927-1931). It would seem that the cases of thermoluminescence with which we are acquainted are really cases of phosphorescence intensified by rise of temperature. The spectrum of thermoluminescent bodies, also, is similar to that of phosphorescent ones. However, not all phosphorescent materials are also thermoluminescent. Living light is quite another phenomenon from thermoluminescence.

Phosphorescence. Although the word phosphorescence has been used in a very loose way to indicate all kinds of luminescence, and particularly that of phosphorus or of luminous animals, to the physicist it has a very definite meaning, namely, the absorption of radiant energy by substances which afterwards give this off as light. If the exciting radiant energy is light (visible or ultraviolet) we speak of photoluminescence, if cathode rays, we have cathodoluminescence, if anode rays, anodoluminescence, and if X-rays (Röntgen rays) we have radioluminescence. Inasmuch as the α, β, and γ rays of radium correspond to anode, cathode, and X-rays, respectively, radium radiation also produces luminescence in many kinds of material. If the material gives off the light for less than a millionth of a second after radiation we can speak of fluorescence. The distinction between phosphorescence and fluorescence is perhaps a purely arbitrary one, as there are a great many bodies which give off light for varying fractions of a second after being illuminated. Some substances also, which fluoresce at ordinary temperatures, will phosphoresce at low temperatures. Phosphorescence is exhibited chiefly by solids but some gases will phosphoresce, notably nitrogen (Lewis, 1900-1914; Strutt, 1911-1935) or mixtures of nitrogen and oxygen on compression (Newall, 1897). The mechanism of this luminescence, which may persist for some time after nitrogen has been subjected to an electric discharge, the so-called active nitrogen, is in all probability due to recombination of nitrogen atoms (Sponer, 1936).

The best known cases of phosphorescence, which occur at room temperature, and the group to which the word phosphorescence is commonly applied, are those of the alkaline earth sulphides, barium, calcium, strontium and zinc sulphides. These are called impurity phosphors and consist of the sulphide, a flux such as sodium sulphate or lithium phosphate and a heavy metal impurity—copper, lead, silver, manganese, zinc, etc. Some pure substances (calcium or zinc oxide) will also phosphoresce but not as brightly. An Italian cobbler, Vicenzo Cascariolo, is said to have discovered the Bologna stone (barium sulphate or barytine) which, by calcination with charcoal, gave an impure phosphorescent barium sulphide, *lapis solaris*, the solar phosphorus or *phosphorus Bononiensis*. The date is not certain, probably 1602 or 1603, for it was only recorded later. Galileo knew about the Bologna phosphorus and showed it to LaGalla in 1612, who published an account. Licetus (1640) gives the history. As early as 1652, Zucchi noted that the color of the light of the phosphor was the same after exposure to sunlight under different colored glasses.

By 1675 three additional phosphors (called "noctilucas" by Boyle) had been discovered (1) Baldouin's phosphorus in 1675, prepared by heating chalk and nitric acid, an impure calcium nitrate, (2) the *phosphorus smaragdinus*, the emerald

TYPES OF LUMINESCENCE

green stone (fluorspar) of Oldenburg which luminesced on warming and (3) *phosphorus fulgurans*, the element phosphorus of Brandt (1669), Kunkel (1674) and Boyle (1680). The light of the element is not a true phosphorescence but a chemiluminescence. Homberg's phosphorus (1693), an impure calcium chloride, was made by heating sal ammoniac and quicklime. It was a grey vitreous substance which glowed on rubbing or striking. John Canton's phosphorus was a calcium sulphide, prepared in 1768, by "*heating a mixture of three parts of siften calcined oyster shells with one part of sulphur to an intense heat for one hour.*" Sidot (1866) discovered that natural zinc sulphide or Sidot blend, which is slightly phosphorescent, may be made markedly so by properly heating it so that the crystals change to the cubic system. On adding copper it gives a brilliant green phosphorescence.

The early workers usually exposed their phosphors to sunlight. Phosphorescence resulting from the ultraviolet light in electric sparks, such as the discharge of a Leyden Jar, was known to Kortum (1794) and Dessaignes (1809) but was probably observed earlier. Dessaignes (1810) observed the phosphorescence (due to cathode rays) of bodies excited by discharges in vacuo, while canal ray effects were seen by Wien (1901) and Schmidt (1902). X-ray phosphorescence was discovered by Winkelmann and Straubel in 1896; that due to radium rays by the Curies, investigated more extensively by H. Becquerel (1899) and Giesel (1899).

The intensity and duration of a phosphorescent light depend chiefly on the nature of the exciting rays, the color chiefly on the impurity present but the flux also exerts an influence. Rise in temperature increases the intensity but diminishes the duration. While relatively few solids give prolonged phosphorescence after exposure to light at ordinary temperature a large number of these acquire the property at the temperature of liquid air. Even pure ice at liquid air temperatures will phosphoresce after exposure to an arc light, as shown by Dewar and by Armstrong (1924). Included also in the list are such biological products as urea, salicylic acid, starch, glue and egg shells. Most of these fluoresce or phosphoresce very briefly at room temperatures.

Special means must be used to observe a phosphorescence of short duration. Beccarius (1734) arranged a special chamber to view objects in the dark immediately after exposure to sunlight and found that nearly everything phosphoresced momentarily, especially paper and dry yolk of egg, but dry egg white did not. E. Becquerel (1867) devised an apparatus for measuring very short phosphorescence, a phosphoroscope. It consists of revolving disks with holes in them between which the object to be examined is placed. The holes are so arranged that the object is first illuminated and then completely cut off from light. The observer looking through another hole sees the material a moment after it has been illuminated and can thus tell if it is phosphorescing. By determining the rate of revolution of the disks it is easy to calculate how long the phosphorescence persists. Many other types of phosphoroscope are known (see Nichols, 1916). A simple one, devised by Andrews (1920), is a wheel on whose periphery the material to be examined is fastened. While revolving, it is illuminated at one place and examined at another. From rate of revolution and length of the persisting luminescent streak the duration of phosphorescence can easily be calculated.

The spectrum of most phosphorescent substances is made up of one or more continuous bands having maxima at different wave lengths (Fig. 71). In the light incident on a phosphorescent substance are also bands of light rays which are absorbed and whose wave lengths are more efficient than

TYPES OF LUMINESCENCE

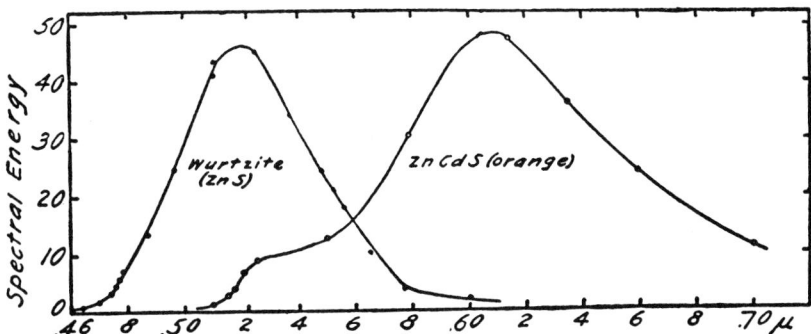

FIG. 71 Spectral energy distribution of some phosphorescences. Wave lengths along horizontal; after Coblentz and Hughes.

others in stimulating phosphorescence. These bands in the phosphorescent light are usually of longer wave length than those in the light which excites the phosphorescence, a fact known as Stokes Law.

The great importance of materials showing phosphorescence and fluorescence for manufacture of television screens and in modern illumination have led to increased investigation of the subject (Curie, 1934; Vanimo, 1935; Rupp, 1937) and interpretation on the basis of excitation of atoms in the crystal lattice (Seitz, 1938; Fonda, 1938, 1939; Johnson, 1939).

Curiously enough, red and infrared rays have the power of annulling phosphorescence by greatly accelerating the rate of decay. After a momentary increase in brightness no more phosphorescence is to be observed. Ives (1910) showed that infrared radiation had no power of quenching the light of the firefly as it does the phosphorescent light of zinc sulphide, an observation cited to demonstrate that the firefly's light is not due to phosphorescence. Many other facts indicate that the light of luminous animals is in no sense a phosphorescence and is quite independent of previous illumination of the animal. In fact, light will often inhibit a bioluminescence.

Fluorescence. The word comes from fluorspar, one of many substances which emit light while they are being radiated but whose light disappears when the radiation is cut off. Liquids and gases, particularly, fluoresce. Usually the emitted radiation is of longer wave lengths than that absorbed, but a gas molecule illuminated with light of a critical wave length may become excited and re-emit the absorbed radiation. This is "resonance radiation." If the excited gas molecule collides with another, transferring its excess energy, the second molecule may emit quite a different wave length, "sensitized fluorescence." If the second molecule does not radiate, the energy goes to heat and we have "quenching of fluorescence." A discussion will be found in Wood's *Physical Optics* (1934).

Fluorescence is readily excited by the cathode rays of a vacuum tube. They not only cause the residual gas in the tube to glow (*electroluminescence*) by which their path may be followed with the eye, but also a vivid fluorescence of the glass walls of the tube, yellow green with sodium glass, blue green with lead and lithium glass. Diamonds, rubies and other minerals fluoresce brilliantly in the path of cathode rays. Fluorescent screens of barium platino-cyanide, willemite (zinc silicate), Sidot blend (zinc sulphide) or scheelite (calcium tungstate) are frequently employed to render visible X-rays. The self-luminous paint most used at the present time is zinc sulphide containing a trace of radium salt whose continually emitted rays cause a steady fluorescence of the zinc sulphide.

Many minerals and most all organic materials examined in ultraviolet light, especially solid or gelled tissues such as bone, teeth, cartilage, lens of the eye, tendon and protein crystals fluoresce beautifully (Fig. 1). Relatively long ultraviolet wave lengths ($\lambda = 0.36\mu$) are most effective. They pass the nickel oxide glass filters which are practically opaque to visible light.

TYPES OF LUMINESCENCE

Another type of fluorescence is that exhibited by solutions. Nicolo Monardes in 1575 refers to the extract of a Spanish wood, good for kidney diseases, which appeared bluish. This *lignum nephriticum*, whose tincture was yellow by transmitted light but bluish from the side, was apparently known to Kircher (1646), Grimaldi (1665), Boyle (1680) and many others later, including Newton (1704), without a realization of the significance of the blue light. Brewster (1833) noted the red beam in a chlorophyll solution when white light was passed through it and called the effect "internal dispersion" thinking it due to colored particles in suspension. Herschel (1845) called a similar blue beam in quinine solutions "epipolarized" but Stokes (1852) was the first to realize that the blue of quinine or the red of chlorophyll was really an emission of blue or red light and not a reflection. He called it at first "true internal dispersion" to distinguish it from "false internal dispersion" or scattering, such as makes a beam of sunlight in a dusty room visible. Later he called it fluorescence and showed that a test tube of quinine sulphate solution held in the ultraviolet region of a solar spectrum would glow with a pale blue light, although it was not illuminated with any rays that are visible to our eyes. Concerning this, Stokes, to whom much of our knowledge of the subject is due, said, "It was certainly a curious sight to see the tube [containing quinine sulphate solution] instantaneously lighted up when plunged into the invisible rays; it was literally 'darkness visible.'" Quinine sulphate absorbs the ultraviolet, converting these rays into a visible blue. Its fluorescence spectrum is a short continuous one. Fluorescent light from solutions is not polarized, while the light scattered from small particles (Tyndall beam) is polarized, an easy method of distinguishing the two phenomena. Many solutions show fluorescence in strong light, both animal and plant extracts. This is especially marked in essential and mineral oils, many colorless organic compounds like esculin or lophin and a long list of aniline dyes—eosin, fluorescein, rhodamin, etc. The green fluorescence of fluoresceine in 10^{-8} gram per cubic centimeter is visible in the beam from an arc lamp.

Minute traces of certain dyes added to a photographic emulsion will make the plate sensitive to green, yellow, red or even infrared rays. They act as photo-sensitizers and may be poisonous to small organisms in light, not in the dark. This type of molecule is frequently characteristic of fluorescent substances, but the fluorescent light itself is not responsible for the photo-sensitization. Sometimes fluorescent substances such as hematoporyphrin, a derivative of haemoglobin, appear in man, creating a light sensitivity that is always aggravating and frequently dangerous.

When a fluorescent dye in solution is irradiated, it may emit light for only 10^{-8} seconds afterwards. If gelatin is added, or if the dye is distributed in films having the general properties of a glass, or if the solution is frozen, the fluorescence is brighter and is converted into a phosphorescence.

Fluorescence microscopes have been devised to study the fluorescence induced within living cells at high magnification (Richert, 1911; Heimstadt, 1911; Lehman, 1913; Singer, 1932). Fluorescence analysis is a recognized method of analytical chemistry (Dankwortt, 1929; Radley and Grant, 1933; Haitinger, 1937), and striking fluorescent substances have been described in animal and plant tissues by Stubel (1911), von Prowazek (1913), Lloyd (1924), Cockayne (1924), Harvey (1926), Singer (1933), Geise and Leighton (1933, 1937), Hamperl (1934), Barnard and Welch (1936) and others. Fig. 72 is a photograph of fluorescence in bacteria. The photogenic glands of luminous animals may fluoresce in ultraviolet light (see Chapter IV) but this behavior does not

TYPES OF LUMINESCENCE

mean that the light of animals is a fluorescence. It only indicates that a compound is present whose molecules can be easily excited to luminesce.

Electroluminescence. This category covers a number of different phenomena in which electric discharges occur, the luminescence of low pressure gas in vacuum tubes and luminescences due to discharges in air. Everyone has observed the luminescence which appears on stroking silk or fur. Merely rubbing a neon tube will cause the gas to glow.

The low pressure discharges were observed as early as 1675 by Picard who noted the light in a Torricellian vacuum at the head of a mercury barometer column when carried from room to room. This was also observed by Bernouilli (1700) and called the "mercurial phosphor" by Hawkesbee (1709). Dufay (1723) wrote a memoir on the subject. The discharge occurs in residual gas from potentials built up when the mercury rolls over the glass. A more refined and instructive example is the tube of low pressure neon containing a drop of mercury. As the mercury rolls from one end of the tube to the other a red discharge occurs where the mercury surface *leaves* the glass. Earrings of such annular tubes with mercury give a pretty display when shaken. What is frequently spoken of as frictional electricity is really electricity resulting from separation of surfaces. Many instances are known. The transient greenish luminescence which occurs at the point where electricians' or surgeons' or "Scotch" tape is stripped from a roll has already been mentioned. The author (1939) has observed that with some samples this luminescence may be so bright that it is visible with only partially dark-adapted eyes. The phenomenon can be repeated if the tape is rewound and then restripped and also appears when the sticky sides of the tape are pressed together and then separated. Rubber cement, whether holding together two pieces of metal, glass, paper, cellophane or any two different materials, luminesces when the surfaces are separated. Many other substances when closely adhering to each other will also luminesce when pulled apart. Films when stripped from glass or metal will give a flash of luminescence; for example, collodion dissolved in ether-alcohol mixtures poured on a glass plate and allowed to dry.

The explanation of all such luminescences appears to be this: whenever two surfaces are separated from each other the capacity diminishes and the voltage rises until a discharge takes place, exciting the surrounding gas to luminesce. That a discharge does actually take place can be readily shown by stripping adhesive tape or Scotch tape in an atmosphere of 2 to 4 cm. Hg pressure of neon gas. Then the luminescence is reddish instead of yellowish. Red luminescence also occurs when two strips of mica are pulled apart or when collodion or ambroid or rubber films are stripped from glass in a low-pressure neon atmosphere. Animal light is in no way connected with these phenomena.

Sonoluminescence. Allied to electroluminescence is the light which appears when intense sound waves pass through fluids, sonoluminescence. This type of light emission was first observed by Frenzel and Schultes (1934) for supersonics and by Chambers (1937) for audible frequencies. Using a magnetostriction oscillator of 8900 frequency, Chambers studied 36 pure liquids and found 14 of them to luminesce. Glycerol and nitrobenzol were especially bright, with water intermediate. Levsin and Kzevkin (1937) found that four-times distilled water will luminesce and conclude that the light is due to electrical discharge in water vapor in cavities within the liquid. Polatskii (1938) and Harvey (1939) have investigated the sonoluminescence in water with various dissolved gases. The author found that bromine water was especially bright and concluded that the light was balloelectrical in origin. All ob-

TYPES OF LUMINESCENCE

servations agree that the luminescence appears when fluids are completely shielded from an electrical field and that it is always connected with cavitation, the formation of gas or vapor cavities in a liquid. The light is brightest where cavitation is most marked.

Balloelectric potentials are produced by increase in the surface of a fluid (as when it is atomized), sometimes called Lenard potentials (Wasserfallelektricität). They have been suggested as an explanation of the extraordinary chemical effects of cavitation on ship's propellors, for they may result in activation of oxygen and corroding chemical reactions. The minute bubbles of gas cavitated by the supersonic waves must develop a considerable electrical charge as their surface increases. When they collapse their capacity decreases and their voltage rises until a discharge takes place in the gas of the bubble, giving rise to a weak luminescence. The discharge may be thought of as occurring at the instant of complete collapse. If oxygen is present in the fluid the discharge leads to activation, with formation of hydrogen peroxide or direct chemical reaction of activated oxygen with certain compounds, such as aminophthalichydrazid, present in solution. A sonic chemiluminescence then results as described by Flosdorf, Chambers and Malisoff (1936).

Galvanoluminescence. Light which appears when solutions are electrolyzed can logically be spoken of as a galvanoluminescence. The light appears at anode or cathode as a result of chemical reaction. Bancroft and Weiser (1914) observed "electrolytic flames," the relatively bright luminescence at mercury anodes when halides, especially sodium bromide, are electrolyzed. They supposed the luminescence to result from combination of halogen and mercury to form the solid salt.

Light at the aluminum anode of an aluminum-carbon rectifier has been known since Braun's (1898) description and was studied by Dufford (1927, 1929), who also observed the luminescence of metal electrodes in ether solutions of Grignard compounds (organo-magnesium halides). A luminescence of the luminous material of animals at a cathode during electrolysis will be considered in Chapter IV, p. 143.

The author has described (1929) an especially bright galvanoluminescence, observed at anodes when alkaline aminophthalichydrazid, the organic compound which can be conveniently called *luminol* (not luminal, the sleeping powder) is electrolyzed. This substance luminesces brilliantly in presence of active oxygen such as appears when hydrogen peroxide or ozone decomposes, or at anodes.

Triboluminescence and Piezoluminescence. Under this head are grouped a number of light phenomena which at first sight may appear to be entirely electrical in nature but are not necessarily so. The light is produced by shaking, rubbing, or crushing crystals, and only crystalline bodies appear to show true triboluminescence or piezoluminescence. Probably Francis Bacon (1620) was the first to record that sugar when crushed would luminesce. Members of the Florence Academy knew of triboluminescence in 1660 and Boyle experimented with diamonds in 1663, finding that they glowed when broken and noting that one of the King's diamonds would shine after "*one brisk stroke of a bodkin.*" He also (1667) knew "*that hard sugar, being nimbly scraped with a knife would afford a sparkling light*" and showed that this took place when sugar was scraped in a vacuum. Probably van Musschenbrock (1667) and certainly Mentzel (1675) observed light on rubbing Bologna phosphorus, while Homberg (1693) prepared his phosphorus, an impure calcium chloride, which glowed with a greenish light when struck. Wedgewood (1792) observed that both sugar and quartz would luminesce when rubbed under water or oil. Even ice is triboluminescent (Precht, 1902).

TYPES OF LUMINESCENCE

A striking case that all can try is that of uranium nitrate. Gentle agitation of the crystals in a glass tube is sufficient to give off sparks of light which much resemble the scintillations of dinoflagellates when sea water containing these animals is agitated. Saccharin crystals will also light if shaken and Pope (1899) found that the bluish light of saccharin was bright enough to be visible in a room in daytime. It only appeared from impure crystals and freshly crystallized specimens. Other crystals, also, have been found to lose their power of lighting after a time. Among biological substances, cane sugar, milk sugar, mannite, hippuric acid, asparagin, r-tartaric acid, l-malic acid, vanillin and benzoic acid are notable. A long list of triboluminescent substances is given by Tschugaeff (1901), by Trautz (1905, 1910), by Gernez (1905) and by Imhof (1917). Trautz found 11.5 per cent of 285 inorganic compounds and 31.5 per cent of 452 organic compounds to be triboluminescent. This excludes the 90 alkaloids studied, of which 63 were triboluminescent. The spectrum is a short continuous one, the waves emitted depending on the kind of crystal (Nelson, 1926). Thus the color of the light varies among different santonin derivatives from yellow to green. In saccharin it is blue.

Triboluminescence can easily be mistaken for electric discharges. Wick (1937) finds that triboluminescent light may result from a number of different causes. It may come from unstable centers in substances which have previously been exposed to light, radium, X-rays or cathode rays. This light would be like a tribophosphorescence, set off by mechanical means instead of by heat as in thermoluminescence. It may also come from centers not dependent on previous radiation and be characteristic of the material itself, true triboluminescence. Excitation would be due to the mechanical energy of breaking. Finally there may be electric discharges in the air surrounding the breaking crystals. In this case the bands of nitrogen are present in the spectrum.

Electric discharges can be easily recognized by shaking or crushing crystals in a low pressure atmosphere of neon. The author has noticed (1939) that practically any crystals and such material as chips of glass, silicious earth or chitin fragments will give a reddish luminescence when shaken in neon. This is electroluminescence. However, in addition to some red flashes, sugar, salophen, salicylamid and uranyl nitrate also show clearly the bright greenish or colorless sparks of true triboluminescence, not definitely reddish in color. Sugar candies made up of a mass of fine sugar crystals luminesce when broken. The "necco" wafer with a wintergreen flavor is especially bright. If broken in a neon atmosphere the same greenish flash occurs as in air but the light is especially bright because fluorescence of the wintergreen oil adds itself to the triboluminescence of the sugar.

Many triboluminescences which occur in the form of scintillations are electrical in nature. Coolidge and Moore (1926) have described these in calcite or celluloid after intense cathode ray bombardment. In addition to the long-lasting orange phosphorescence of calcite, bluish-white scintillations were observed due to electric discharges, minute explosions which left a crater with many tiny radiating canals. They could be induced by scratching the surface of the crystal. The electrons shot into the surface layers of the dielectric build up high potential gradients which discharge disruptively.

Although the light produced by some living organisms resembles triboluminescence in that it may be evoked by rubbing or shaking the animals, it is in reality fundamentally different since it is dependent on the presence of oxygen whereas triboluminescence is not.

TYPES OF LUMINESCENCE

Crystalloluminescence and Lyoluminescence. Crystalloluminescence is observed when solutions crystallize. It was described first for potassium sulphate by Schönwald (1786); also by Pickel (1787), Giobert (1790) and Schiller (1791). Between 1823 and 1835 other crystalloluminescences became known and Rose (1835) added arsenious acid to the list. Pontus (1833) had observed a vivid spark of light when water in a small glass globe was quickly frozen. All well authenticated cases of crystalloluminescence are exhibited by simple inorganic salts and these are also all triboluminescent. The reverse is not true, however, many triboluminescent substances are not crystalloluminescent. Crystalloluminescence is much less widespread than triboluminescence. Trautz and Schorigin (1905) have studied the matter in a number of compounds and come to the conclusion that the light is really a special case of triboluminescence in which the growth of individual crystals causes them to rub together. The light becomes much brighter on stirring a mass of crystals which are forming. While in some cases crystalloluminescence is unquestionably due to the triboluminescence of crystals rubbing against each other, it is not in every case, as is indicated by the work of Weiser (1918). He studied luminescence of saturated aqueous alkali halide solutions (sodium or potassium chloride) upon addition of alcohol or hydrochloric acid. The salt crystallizes out rapidly under these conditions and Weiser found that the light is brightest when the conditions of concentration of alcohol or of acid are such as to cause heaping up of sodium and chloride ions. He believes that the bluish light which appears is due to the combination of ions in the reaction, $Na^+ + Cl^- = NaCl$. Only if this proceeds rapidly enough does luminescence occur. Weiser studied also the crystalloluminescence and triboluminescence of arsenious chloride and of potassium sulphate. By photographing the luminescence through color screens

(Weiser, 1918), a spectrum of the light could be obtained, and it was found to be identical in both the tribo- and crystalloluminescent light. Weiser believes the light in this case also to come from recombination of the ions, $As^{+++} + 3Cl^- = AsCl_3$, and that crystalloluminescence in general is due to rapid reformation of molecules from ions broken up by electrolytic dissociation, while triboluminescence is due to rapid reformation of molecules from ions broken up by violent disruption of the crystal. Of course in triboluminescent organic crystals which do not dissociate into ions, some other reaction must be responsible.

Crystals are not found in the luminous organs of animals with the exception of the fireflies. In these a layer of cells occurs (see Chapter II) filled with minute crystals of one of the purine bases (xanthin or uric acid). One might surmise that the light of the animal was a crystalloluminescence accompanying the formation of these crystals. It is easy to show, however, that the light comes not from the crystal layer but from another layer of cells containing large granules. It is also dependent on the presence of oxygen while crystalloluminescence takes place in the absence of oxygen.

The light of luminous organisms is quite generally associated with granules. In one of the centipedes (*Orya barbarica*) which produces a luminous secretion, Dubois (1893) has described the transformation of these granules into crystals and at one time he supposed the light to be a crystalloluminescence but later reversed this opinion.

The phenomenon of *lyoluminescence*, solution luminescence, was described by Wiedemann and Schmidt (1895) as a light accompanying the solution of colored (from exposure to cathode rays) crystals of lithium, sodium or potassium chlorides. Schwarz (1903) also described luminescence during solution of sugar, brighter the higher the temperature, i.e. the

TYPES OF LUMINESCENCE

more rapidly solution occurs. This is possibly a triboluminescence.

Chemiluminescence. As the name implies, chemiluminescence is the production of light during a chemical reaction at low temperatures. This does not mean that other types of luminescences are not connected with chemical reactions, using the word *reaction* in a broad sense. Garrison (1927) speaks of fluorescence and phosphorescence as indirect chemiluminescences, since the changes may be chemical. However, the energy for the change comes from the exciting radiation. The direct chemiluminescences result when radiant energy appears at the expense of chemical energy. The general mechanism of such transfers has been discussed by Taylor (1927). They might include the luminescence of flames (pyroluminescence) already described, in which the spectra are not characteristic of the element as such but rather of a particular reaction in which the element takes part (dissociation into ions, changes from monovalent to bivalent condition, etc.). This is the reason one element may show various spectra under different conditions (Bancroft, 1913). Highly diluted flames of sodium and halides have been extensively investigated by Polanyi and collaborators (1928).

Direct chemiluminescences also include reactions in which chemical splitting of the molecule occurs (as contrasted with ion reactions), chiefly oxidations. True chemiluminescences are oxidation reactions involving valence changes or the absorption of gaseous or dissolved oxygen and may be very easily distinguished from all the previously mentioned luminescences by this criterion (see Weiss, 1938). They should, perhaps, more properly be called *oxyluminescences*. The oxidation may occur in the gas or liquid phase. The glow of phosphorus is the best known case. A cold flame, below room temperature, can be easily obtained by boiling a solution of phosphorus in chloroform under reduced pressure, and mixing the vapors with air. Much study has been devoted to the luminescence of this element and some interesting periodic light phenomena described at low oxygen pressures (Ewan, 1895; Centnerswer, 1895; Russel, 1903; Scharff, 1908; Bloch, 1908-1909; Rayleigh, 1921; Weiser and Garrison, 1921; Petrikaln, 1924).

Davy noticed that fresh cut surfaces of sodium and potassium metal will glow in the dark for some time, especially if warmed to 60°-70°C. (Linnemann, 1858; Reboul, 1919). A film of oxide is formed over the surface, showing definitely that oxidation has occurred.

Ozone oxidizes organic matter with an accompanying glow as shown by many observers (see Stuchthey, 1920; Trautz and Seidell, 1922). The author has observed (1917) that the light from ozone bubbled through dilute pyrogallol solution is especially bright under certain conditions.

Radziszewski (1877, 1880, 1883) gives a long list of substances, chiefly essential oils, which luminesce if slowly oxidized in alcoholic solutions of alkalis. Formaldehyde, dioxymethylen, paraldehyde, metaldehyde, acroleïn, aldehydeammonia, acrylammonia, hydrobenzamid, lophin, hydroanisamid, anisidin, hydrocuminamid, hydrocinamid, besides waxes and such biological substances as glucose, lecithin, cholesterin, cholic, taurocholic, and glycocholic acids and cerebrin, all luminesce on oxidation. Lophin and its derivatives are especially well known. Radziszewski himself and many other authors have compared the light of organisms to this type of luminescence. Indeed the incorrect identification of granules found in the cells of practically all luminous tissues as oil droplets, is largely due to the influence of Radziszewski's work.

Dubois (1901) has added esculin, and Trautz (1904-1905) many aldehydes and phenol derivatives, including vanillin, papaverin, tannic and gallic acids, besides glycerol and mannite

TYPES OF LUMINESCENCE

to the list of biological substances oxidizing with light production. Guinchant (1905) has described oxyluminescence of uric acid and asparagine, Weitlaner (1911) of substances in humus and McDermott (1913) of substances in urine and the anaerobic alkaline hydrolysis products of glue and Witte's peptone.

Pyrogallol is especially prone to luminesce, as was first noticed by Eder (1887) and Lenard and Wolf (1888) in developing a photographic plate with pyrogallol developer. Many articles have appeared from time to time describing the luminescence of photographic plates. Later the luminescence was studied in some detail by Trautz and Schorigin (1904-1905), who developed the well known luminescent mixture of pyrogallol, formaldehyde, potassium carbonate, and hydrogen peroxide. As both Goss (1917) and the author (1916, 1917) have shown, pyrogallol can be oxidized in many different ways. Some of these are of great interest, for they very closely imitate the mechanism for the production of light in organisms.

Another group of chemiluminescences, described by Delépine (1910-1912), is connected with the oxidation of sulphur compounds—thiophosgene, esters of thio analogues of carbonic acid and organic derivatives of the thiophosphoroso radical combined with chlorine. The S-C-O structure is frequently found in luminescent sulphur compounds.

The chemiluminescence of a purely inorganic silicon compound, siloxene ($Si_6H_6O_3$), is of special interest since its study has thrown much light on the mechanism of luminescence and especially the relation between chemiluminescence and photoluminescence (Kautsky, 1921; Kautsky and Zocher, 1922; Kautsky and Thiele, 1925; Kautsky and Neitzke, 1925; Kautsky and Hohn, 1936). When siloxene is oxidized by permanganate, nitric acid, hydrogen peroxide and other strong oxidants, a reddish luminescence appears which is attributed to the excitation of other unoxidized siloxene molecules. When exposed to ultraviolet light, a similar reddish luminescence (fluorescence) is observed due to excitation of siloxene molecules by the energy of the ultraviolet light. The two spectra have the same maximum and in each case the light is polarized. Since the fluorescence occurs at liquid air temperatures, where chemical reactions proceed with negligible velocity, we have strong proof that the luminescence in ultraviolet light is not due to a photochemical reaction but to direct excitation of molecules.

Transfer of the energy of excited siloxene molecules to other compounds, which then luminesce, can also be observed. If rhodamine-B or eosin are adsorbed on siloxene, which has a permutoid structure, and the siloxene slowly oxidized, a chemiluminescence of the rhodamine-B, yellow in color, or of the eosin, green in color, appears. That this effect is not due to fluorescence of these dyes from the chemiluminescent light of siloxene can be shown in two ways: first, by the fact that only eosin or rhodamine-B molecules actually in contact with siloxene surfaces emit light, and second, because the effect is obtained when the dyes are adsorbed on siloxene so completely oxidized that its own molecules do not emit enough light to be visible. We shall see in Chapter IV that at least one bioluminescence occurs as a result of transfer of energy from one molecule to another, which becomes excited and emits the light.

A number of organic metal compounds will luminesce. A recently described instance (Helberger, 1938) is magnesium phthalocyanin, as well as magnesium and zinc complexes of tetrabenzoporyphyrin, phaeophorphyrin or methylchlorophyllid. When these compounds are dissolved in tetralin and hydrogen peroxide or benzoyl peroxide added, a beautiful red luminescence appears, similar to the red fluorescence of chlorophyll.

TYPES OF LUMINESCENCE

The Grignard compounds, which are organo-magnesium halides, such as phenylmagnesium iodide, will luminesce in ether solution in air or oxygen. One of the earliest observations (Wedekind, 1906) was the luminescence of this compound with chlorpicrin, a reaction which does not appear to involve oxidation, although it is possible that it may be more complicated than usually represented:

$$3\ C_6H_5MgI + Cl_3CNO_2 = (C_6H_5)_3CNO_2 + 3\ MgICl$$

A large number of substitution products have been studied by Heczko (1911), Schmidlin (1912), Moeller (1914), Lifschitz and Kalberer (1922), and especially by Dufford and collaborators (1923-1933) in an attempt to correlate brightness of luminescence with some particular type of structure. Para-chlorphenylmagnesium bromide is especially bright, although Backman (1934) obtained a striking luminescence with 9-phenanthrylmagnesium bromide, and Hill[4] with β-naphthylmagnesium bromide. Intermediate products remain unknown so that it is impossible to suggest the mechanism of luminescence or the excited molecule. Grignard compounds are weakly fluorescent, their oxidation products strongly so, in ultraviolet light. Since the oxidation occurs in ether solutions, the light can have little connection with bioluminescence.

In 1928 Albrecht published a thesis dealing with the brilliant blue luminescence of alkaline solutions of aminophthalichydrazid or luminol, a compound known since 1902 (Schmitz) but whose luminescence was first observed by Dr. W. Lommel, in Leverkusen. A number of derivatives were investigated, as well as the conditions for luminescence, especially the light which appears in presence of active oxygen from catalytic decomposition of hydrogen peroxide, or from ozone, or upon addition of ferricyanides, hypochlorites, and many other oxidants. The light from one part luminol in 10^8 parts

[4] Private communication from A. J. Hill (1929).

water is visible. Luminol gives a bright blue fluorescence in neutral or acid but not in alkaline solution.

Since Albrecht's paper, numerous workers (Harris and Parker, 1935; Witte, 1935; Gleu and Pfannsteil, 1936; Zellner and Dougherty, 1937; Drew and collaborators, 1937-1938; Thielert and Pfeiffer, 1938; Svesnikov, 1938; Schales, 1939) have continued the investigation. A striking demonstration may be made by soaking a white towel in alkaline luminol containing some hydrogen peroxide and pouring potassium ferricyanide solution over it (Huntress, Stanley and Parker, 1934). The towel glows like a live coal and on wringing yields liquid drops of cold fire.

Reference has already been made to the luminescence of this compound at anodes and as a test for active oxygen in sonic effects. It will luminesce in contact with phosphorus, when oxy-gas flames are directed on the surface of its solution, or when in contact with metals like aluminum, zinc, cadmium or tin. In all these cases hydrogen peroxide is formed. Even the metals decompose water with liberation of hydrogen, which forms hydrogen peroxide with the surrounding oxygen. Decomposition of the hydrogen peroxide thus liberates atomic oxygen, which is responsible for chemiluminescence of the luminol. Hydrogen peroxide alone added to luminol gives only a faint luminescence; hemoglobin or plant peroxidases alone gives none[5] but a combination of the two results in a brilliant light.

The author has observed (1929) that oxidation with bright luminescence can occur if luminol and ferricyanide are mixed in an atmosphere of pure hydrogen. No active oxygen is involved but an electron transfer, similar to the oxidation of hydroquinone. Various interpretations of the mechanism of

[5] Tamamushi (1937) has observed luminescence of luminol and haemine with pure oxygen.

TYPES OF LUMINESCENCE

light production have been made but little more than provisional schemes can be suggested (see Drew, 1938).

Another group of compounds, likewise giving a brilliant chemiluminescence but of a greenish color, are the quartenary salts of dimethyldiacridylium. Gleu and Petsch (1935) showed that oxidation with long lasting luminescence occurs in alkaline solution with hydrogen peroxide. A little osmium tetroxide added to this catalyses decomposition of hydrogen peroxide and enormously increases the light intensity.

In acid or neutral solution of the nitrate, a strong greenish fluorescence is apparent but none in alkaline solution. The compounds have been called "lucigenines" and are said to be 100 times as bright as luminol (Decker and Petsch, 1935). One part in 10^{10} parts water can easily be recognized.

Contrary to the conditions for luminol light, hypochlorite or ferricyanide do not cause luminescence, but reducing agents such as hydrosulphites, sulphides and stannites in alkaline solution do, provided oxygen is present. If the oxygen is removed by a stream of nitrogen the luminescence disappears, to reappear if oxygen is again admitted. The light is apparently connected with the last of the following series of reactions (1) carbinol base formation by alkali (2) peroxide formation by oxidation with oxygen or hydrogen peroxide (3) reduction of the lucigenine peroxide to a carbinol base. Hydrogen peroxide plays an extraordinary dual rôle—acting first as an oxidizing and then as a reducing agent. Sulphides or stannites act by reducing the lucigenine peroxide previously formed on oxidation with dissolved oxygen. Light accompanies the reduction.

Space does not permit a consideration of the spectra of chemiluminescent vapors or of gases, such as the afterglow of nitrogen, despite the importance for an analysis of the mechanism of light production. In some reactions, such as the decomposition of ozone, light may be emitted in the ultraviolet; in

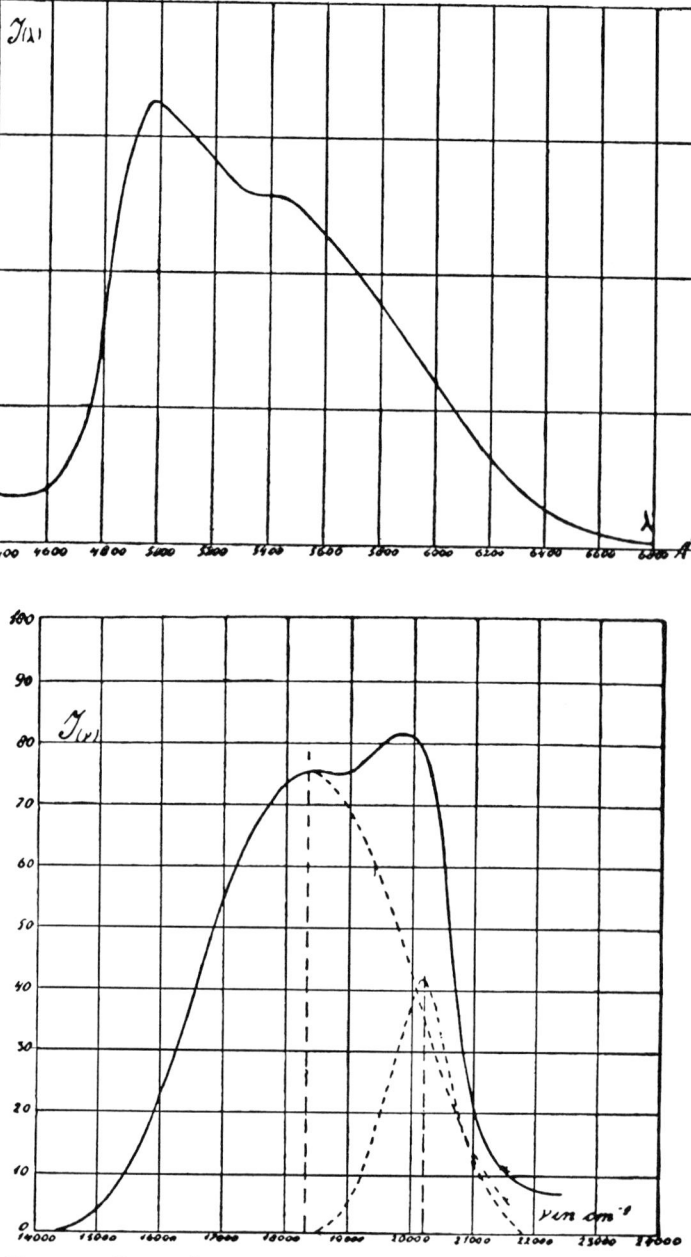

FIG. 73 Spectral energy distribution of the chemiluminescence of dimethyldiacridylium nitrate, plotted both as a function of wave length, λ (above) and frequency, cm^{-1} (below). Note that certain fundamental frequencies are emphasized in the frequency plot. After Eymers and van Schouwenburg.

TYPES OF LUMINESCENCE

the case of glowing phosphorus very short radiation is produced, as far as $\lambda = 0.12\mu$ (Centnerszwer and Petrikaln, 1912; Downey, 1924).

The earliest attempts to study chemiluminescent spectra in solution were made by Radziszewski (1880) who found the light of lophin oxidized in alcoholic caustic alkali, to have a continuous spectrum, brightest at E, with the red and violet ends lacking. Trautz (1905) observed that the pyrogallol-formaldehyde-carbohydrate-peroxide reaction gave a continuous spectrum from the red to the blue green with maximum brightness in the orange. Weak chemiluminescence spectra in solution can be studied by photographing the light through a series of colored filters, without the use of a prism (Weiser, 1918). In this way the maxima for pyrogallol, phosphorus and amarin luminescence were found to lie in different regions. With long exposures, a prism can be used and the spectra of Grignard compound chemiluminescences have been photographed by Dufford, Nightingale and Calvert (1925) and Evans and Diepenhorst (1926); siloxene luminescence by Kautsky and Neitze (1925); luminol by Albrecht (1928) and amarin by Bhatnagar and Mathur (1932). They are all continuous bands in various regions of the visible. The spectral energy curves of luminol and dimethyldiacrylidium nitrate (Fig. 73) have been analyzed by Eymers and van Schouwenburg (1936) and the significance of certain fundamental frequencies pointed out. They will be considered in Chapter VI.

Bioluminescence. No fact speaks against the view that living light is another example of chemiluminescence in which special compounds manufactured by luminous animals are oxidized with light production. It is quite certain that none of the substances already mentioned as capable of luminescence are used in the living world. What these substances are and the history of their discovery will be presented in the next chapter.

CHAPTER IV

THE CHEMISTRY OF LIGHT PRODUCTION

BIOCHEMICAL progress has been marked by a series of sensational syntheses. An epoch started when urea, excretory product in urine, was artificially prepared in 1828. Another, a commercial period, came with the synthesis in 1870 of alizarin, the madder pigment used by the Egyptians to dye their mummy cloths, quickly followed by muscarin (1876), an alkaloid from toadstools. The hormone, adrenaline, was prepared in 1904 and the vitamine, ascorbic acid, in 1933. Enzymes and proteins can be crystallized but not synthesized; luminous substances have neither been crystallized nor synthesized but undoubtedly will be in time. The first step is to isolate and purify them. Considerable progress has already been made along these lines.

LUMINESCENCE AND WATER

Early workers on animal light sought to find a means of preserving the luminescence. Réaumur (1723) was the first to dry the luminous slime of *Pholas* and found that it would again luminesce if water was added. Spallanzani's (1794) results with the medusa, *Pelagia*, were the same and practically every form, fireflies, *Pyrosoma*, *Phyllirrhoë*, myriapods, fungi, copepods, ostracods, pennatulids, even luminous bacteria, if dried quickly, and not kept too long, will again give light on moistening. The ostracod, *Cypridina*, when dried, will retain its power of luminescence practically indefinitely. The author has kept some for twenty-one years, which are still brilliant when moistened. It is possible that the relatively quick loss of photogenic

CHEMISTRY OF LUMINESCENCE

power with dried luminous bacteria is merely an indication that they contain very little photogenic substance.

These experiments show also that animal luminescence is not a *vital* process, in the same sense that the conduction of a nerve impulse is a vital process. A nerve loses its characteristic property of conduction on drying or maceration while luminous cells still possess the power to luminesce after such treatments. Using the terminology of the older physiology, we may say that "living protoplasm" is not necessary for light production.

LUMINESCENCE AND OXYGEN

The necessity of oxygen for luminescence was also an early discovery, proven by the air pump experiments of Boyle (1667) with shining wood and flesh, already described. Since that time every luminous organism, with a few exceptions, has been found to be dependent on free oxygen for luminescence.

It should be clearly borne in mind that if we place luminous organisms, say bacteria or fungi, in an atmosphere devoid of oxygen and find that no light is produced, this may merely mean that certain functions of the cell are interfered with, including light production, but does not necessarily indicate that oxygen is actually used up in the photogenic process. If we find, however, that extracts of luminous cells or luminous secretions devoid of cells cease to give light when the oxygen is removed and again luminesce when it is returned, we may be quite certain that the photogenic process itself requires free oxygen. As luminous extracts of marine worms, earthworms, fireflies, brittle stars, pennatulids, copepods, ostracods, *Pholas*, balanoglossids and many other animals give off no light when the oxygen is removed, we may safely conclude that for these luminescences, oxygen is necessary. Bacteria, fungi, and *Noctiluca*, whose light disappears in absence of oxygen, although they are whole cells, must also by analogy use oxygen in the photogenic process.

Some of the earlier workers on fireflies (Carradori, 1797; Macartney, 1810) and later ones on *Noctiluca* (Massart, 1893) and *Cypridina* (Kanda, 1920) obtained light, supposedly after placing these organs in absence of oxygen, but they did not realize how low is the amount of oxygen necessary for light production. It is difficult to remove traces of oxygen from the water, traces which are nevertheless sufficient to cause luminescence (Harvey, 1920, 1926).

Figures on the minimal concentration of oxygen for visible luminescence of bacteria can now be given. Harvey and Morrison (1923) found that the light was just visible when hydrogen gas containing only one part of oxygen in 100,000 was bubbled through a dilute suspension of marine luminous bacteria. More concentrated suspensions can use the small amount of oxygen so quickly that no luminescence appears. The size of the vessel and the variation in minimum light that will affect

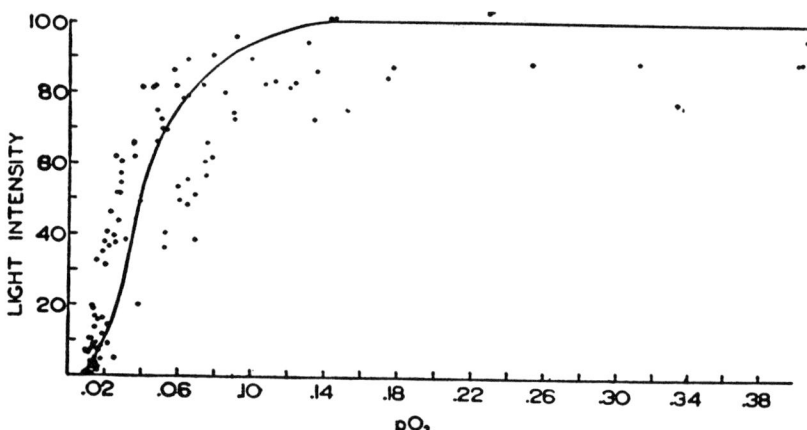

FIG. 74 Relation of the luminescence intensity of a fresh water bacterium (*Vibrio phosphorescens*) to oxygen pressure, pO_2, expressed in per cent of an atmosphere. After Shapiro.

dark adapted eyes make it difficult to set an exact figure for minimum detectable oxygen tension, but we may place it at about 0.0007 mm. Hg. A solution of the luminous substances of *Cypridina* still glows at oxygen pressures which cannot be detected by bacteria, but exact figures are not available.

Shapiro (1934) has studied, by a photocell method, the relation of luminescence intensity to oxygen tension for the fresh-water *Vibrio phosphorescens* (Fig. 74). The light is constant between the 21 per cent of oxygen in air and 0.14 per cent oxygen and then decreases, falling to half intensity at 0.02 per cent oxygen. Light could no longer be detected by the photocell with one part oxygen in 10,000. It is interesting to compare this curve with the curve for total oxygen consumption at different oxygen tensions as determined by Shoup (1929) for a marine form (Fig. 75). Total oxygen consumption was con-

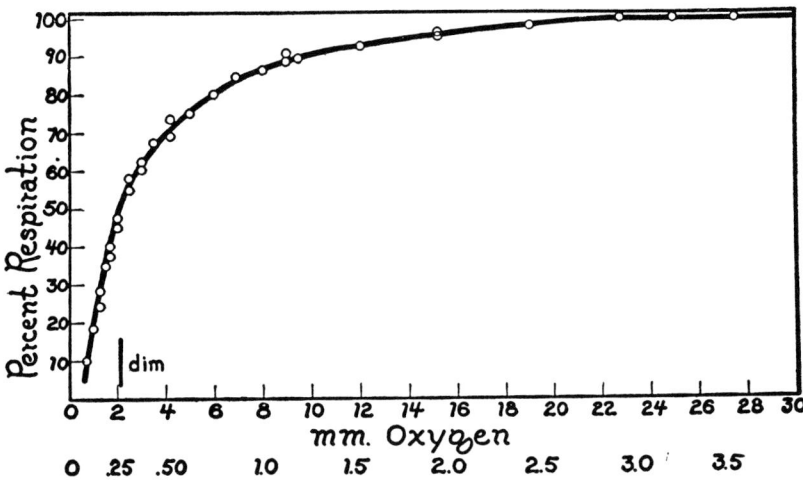

FIG. 75 Relation of the total oxygen consumption of a marine luminous bacterium to oxygen pressure, expressed as mm Hg and as per cent of an atmosphere (lower figures). At the pressure marked "dim," 0.25 per cent, the light intensity of these bacteria first begins to be diminished. Note that at this pressure the oxygen consumption (respiration) is reduced 50 per cent. After Shoup.

stant from the 21 per cent of oxygen in air to 3 per cent oxygen and was reduced to half at 0.25 per cent oxygen. At this point the light intensity just begins to be affected. The independence over a wide range of oxygen tensions of both luminescence and respiration has been confirmed by van Schouwenburg (1938). In pure oxygen a decrease in the respiration of luminous bacteria has been found by both Shoup (1930) and Clarens (1938).

If luminous bacteria are deprived of oxygen, some species (*Photobacterium phosphoreum*) do not luminesce brightly afterwards when air is admitted, while other species (*Ph. fischeri*) are not affected by relatively long periods of anaerobiosis. All forms give a "flash" at the time oxygen is supplied to a suspension which has become dim or dark from an insufficient oxygen supply. The author (1932) first recorded (Fig. 69) this "excess luminescence" and showed that, within limits, the amount of light in the flash was independent of the duration of anaerobiosis. The total light in the flash was interpreted as a measure of the amount of photogenic material stored up when its continuous oxidation is prevented.

Recently Johnson, van Schouwenburg and van der Burg (1939) have made a careful investigation of the flash in relation to duration of anaerobiosis and supply of food material, as well as to cyanide and urethane poisoning (see p. 172 and p. 190). It was found that only after very long periods of anaerobiosis without fermentable substrate was the flash greatly reduced, apparently because of proteolytic destruction of luciferase. This proteolysis was slowly reversed in presence of oxygen. Most significant was the use of the flash method for study of the formation and oxidation of luciferin. If well washed and aerated bacteria, containing no food material, were deprived of oxygen, they gave only a slight flash when it was readmitted; but if glucose was first supplied *anaerobically* and then oxygen admitted, the total light in the flash was vastly greater. This experiment shows that anaerobic metabolism of glucose leads to the accumulation of luciferin. Other experiments indicated that the glucose is not directly transformed into luciferin but acts as a hydrogen donator to reduce the oxidized luciferin already stored up in the bacteria, according to the scheme presented on p. 157.

In a few animals, luminescence can occur when special precautions are taken to remove the last trace of oxygen. Extracts of radiolaria, ctenophores, the medusa, *Pelagia*, (Harvey, 1926, Harvey and Korr, 1938) as well as the medusa, Aequorea,[1] from Puget Sound luminesce in presence of hydrogen and platinized asbestos or of sodium hydrosulphite ($Na_2S_2O_4$), both active absorbers of dissolved oxygen. We are led to the conclusion, either that dissolved oxygen is not necessary or that oxygen is already bound with the photogenic material. The latter alternative seems more probable although chemiluminescence has been observed with other reactions than those using molecular oxygen.

PRODUCTION OF CARBON DIOXIDE

Boyle made many experiments to show that air was necessary for the life of animals and the germination of seeds and found that repeatedly respired air was unfit for further breathing. About the same time Hooke discovered the true meaning of respiratory movements. By forcing a blast of air continuously through the lungs with bellows, he was able to keep dying animals alive. Thus, combustion, respiration and luminescence of flesh or wood were early recognized as related phenomena.

Later it was realized that the slow combustion did not take place in the lungs, or in the blood, but in the tissue cells themselves, and respiration in the chemical sense has come to mean

[1] Private communication from Dr. R. S. Anderson.

this universal slow combustion in the cells of the body rather than the actual breathing movements of the lungs. In anaerobic respiration, carbon dioxide is given off, but no oxygen is absorbed. In aerobic respiration, oxygen is absorbed and carbon dioxide formed, but many substances can oxidize by taking up oxygen without giving off carbon dioxide. It is important to know whether carbon dioxide appears during luminescence and to understand the relation between respiration and light production.

To determine if carbon dioxide is formed during luminescence it is necessary to work with fairly pure luminous materials, obtained from luminous organisms. It is impossible to use living organisms themselves as the carbon dioxide continually respired is a very disturbing factor. If crude extracts of *Cypridina* are used, it can be shown that not enough carbon dioxide is produced during luminescence to change the color of phenol-sulphone-phthalein (Harvey, 1919). Even if a bright flash of light appears, no color change of the indicator is to be observed. Since these solutions contain protein and other buffers and the concentration of luciferin is unknown, a statement of the carbon dioxide detectable by this method cannot be given.

Reactions involving the production of carbon dioxide evolve considerable heat. Studies of heat production during luminescence of *Cypridina* show that the rise in temperature is less than that from the heat produced by stirring and mixing the solutions, certainly less than $0.001°C.$ rise per gram of solution (1919). That animals emit "cold light" is amply verified. However, it is quite certain that some heat is produced. Most luminescence is light without heat only in the sense that very little heat appears as compared with the great amount from incandescence. In both cases some kind of energy is converted

CHEMISTRY OF LUMINESCENCE

into light energy and in the transformation, heat *must* appear. Cold light is only relatively cold.

Since luminescences are easily visible with minute amounts of oxidizing substances in solution, these carbon dioxide and heat production experiments may merely mean that the methods are not sensitive enough to detect the heat and carbon dioxide formed. Other considerations, however, make it seem very unlikely (see p. 155) that carbon dioxide is produced but the final proof must await the isolation of the pure photogen.

PHOTOGEN

When the dried powdered luminous material of an ostracod is sprinkled over the surface of water, it goes into solution and leaves luminous diffusion and convection trails plainly visible in the water. Many luminous marine forms give off a phosphorescent slime which adheres to the fingers when they are handled. It is not surprising that this luminous matter should have early received a name. In 1872, Phipson called it *noctilucine* and described some of its properties. He regarded the luminous matter which can be scraped from dead fish (luminous bacteria) and the mucous secretion of centipedes and the luminous matter of the glowworm, to be this material, "noctilucine," which, "in moist condition, takes up oxygen and gives off carbon dioxide and when dry appears like mucin." Phipson says that it forms an oily layer over the seas in summer (he probably refers to masses of dinoflagellates), is liquid at ordinary temperatures and less dense than water, smells a little like caprylic acid, is insoluble in water but miscible with it, insoluble in alcohol and ether, dissolves with decomposition in mineral acids and alkalies and contains no phosphorus. We can see from this description that the word "noctilucine" does not indicate any chemical individual, but it is the earliest attempt definitely to designate the luminous substance.

LIVING LIGHT

The idea of a substance oxidizing and causing the light has been upheld by a number of investigators, and later Molisch (1904, 1912) called this compound in bacteria the *photogen*. Its formation depended on the life of the cell but it luminesced independently in presence of water and oxygen. He contrasts the "photogen theory" with certain other views of light production which may be spoken of as "vital theories." Pflüger (1875) looked upon luminescence as a property of living protoplasm since light appeared on stimulation. It was a sign of intense respiration, which in most cells produced heat but no light. Sachs (1887) also thought in terms of "Phosphoreszenz durch Atmung." Beijerinck (1891) regarded the light as an accompaniment of the formation of living matter from peptone, but later (1915) thought of the material formed as "photoplasma" which behaved like an endoenzyme and reacted with oxygen to give carbon dioxide.

It was therefore early recognized that water, oxygen and a photogenic substance were necessary for light production. The substance was at first identified with the well known phosphors which absorb light and then emit it, or with phosphorus dissolved in the fluids of the animal. Panceri (1872) thought the material was oil and after Radziszewski's (1877, 1880) studies on chemiluminescence of organic compounds, mostly oils, the incorrect idea that the photogen was an oil became a permanent fixture in bioluminescence literature.

LUCIFERIN AND LUCIFERASE

A very great advance in our knowledge of the chemistry of the problem was made by Dubois in 1885. He showed that if one dips the luminous organ of *Pyrophorus* in hot water, the light disappears and will not return again; also if one grinds up a luminous organ, the mass will glow for some time, but the light finally disappears. When the previously heated organ is brought in contact with the unheated triturated organ, it will again give off light. Later, Dubois (1887) found that the same experiment could be performed with the luminous tissues of *Pholas dactylus*. A hot-water extract of the luminous tissue allowed to cool, and a cold-water extract of the luminous tissue, allowed to stand until the light disappears, will again produce light if mixed together. Dubois advanced the theory that in the hot-water extract there is a substance, *luciférine*, not destroyed by heating, which oxidizes with light production in the presence of an enzyme, *luciférase*, which is destroyed on heating. The luciferase is present together with luciferin in the cold-water extract, but the luciferin is soon oxidized and luciferase alone remains. Mixing a solution of luciferin and luciferase always results in light production until the luciferin is again oxidized. In 1916 I found similar substances in the American fireflies, *Photinus* and *Photuris*, the Japanese firefly, *Luciola*, and in the ostracod crustacean, *Cypridina hilgendorfii*.

This last animal has become a classic in chemical investigation since the stability of the material, the brightness of the light and the ease of preparation of luciferin and luciferase place it far above other forms. In *Cypridina*, the luciferin and luciferase occur as granules in the gland cells but they immediately dissolve in sea water and light comes from granule-free solutions.

In early work the author noticed (1916, 1917) that adding sodium chloride crystals, which could not possibly be oxidized, to a crude luciferase solution, caused luminescence. It was also impossible to oxidize the thermostable component of *Cypridina* with luminescence[2] by permanganate or hydrogen peroxide, in contrast with Dubois's results from *Pholas* material (see p. 135). Consequently, I considered the heat-resistant component of *Cypridina* to be an accessory body,

[2] Although it will oxidize without luminescence.

comparable to co-zymase in alcoholic fermentation of glucose, and called it "photophelein" (from *phos*, light and *opheleo*, to assist). The heat sensitive substance was called "photogenin" (from *phos*, light and *gennao*, to produce), comparable to zymase proper, and regarded as the source of the light. Although photogenin does supply the molecules which emit the light, the author long ago abandoned the view that the system of substances concerned in light production is similar to the zymase—co-zymase system of yeast, and has adopted Dubois's term, luciferase (=photogenin) for the thermolabile material, and luciferin (=photophelein) for the thermostable material.[3] The salt effect is now known to be characteristic of this luminescent reaction, increasing the total light emitted (Anderson, 1937). Addition of salt to a newly prepared crude luciferase solution in which no light is visible, can increase the luminescence intensity to the point where it is visible.

PHOTOPHELEIN

Many instances of granules in luminous gland cells have been already given. The name photophelein is useful to designate a group of substances whose action on photogenic granules is similar to that of cytolytic substances on cells. We may think of the granules as surrounded by a film similar to that at the cell surface. Luminescence accompanies the breakdown of granules and in some cases this process is superposed on the luciferin-luciferase reaction. For example, the slime of *Pholas*, allowed to stand until its light disappears, will give light if mixed with cytolytic agents, such as saponin or ether, or on slow heating or addition of water; also on adding sea-water extracts of non-luminous forms such as *Mytilus*, *Pholadidea*, or non-luminous parts of *Pholas* itself. These are all cytolytic ef-

[3] This does not preclude the possibility that a coenzyme may not also be necessary for the action of luciferase on luciferin.

CHEMISTRY OF LUMINESCENCE

fects, the solution of granules with light emission, granulolysis, which can be observed under the microscope. Pholas luciferase is still present in the solutions and will give a brighter light if Pholas luciferin is added—the true luciferin-luciferase reaction. The substances in the non-luminous animals or in the non-luminous parts of Pholas which cause dissolution of granules can be called photopheleins and are akin to cytolysins. In non-luminous parts of the firefly also, and in non-luminous insects there are photopheleins which will cause solution of firefly luciferase with luminescence.

In Cypridina, light from solution of granules in the extract is a minor phenomenon, since Cypridina granules dissolve readily on contact with sea water, but can confuse the interpretation of luminescence. Extracts of non-luminous animals will cause luminescence of a concentrated crude extract containing luciferase. This is again due to solution of granules, and does not appear if the luciferase is first treated with cytolytic agents such as chloroform.

SPECIFICITY OF PHOTOGENIC SUBSTANCES

Since the discovery of luciferin and luciferase in Cypridina, I have tested many different groups of luminous forms (1922, 1926, 1931). In only five out of twenty-one can the luciferin-luciferase reaction be demonstrated, despite the fact that luminous material is abundant. Hickling (1925, 1926) reports the reaction in the fish, Malacocephalus laevis, but Haneda (1938), as we have seen in Chapter II, was able to grow luminous bacterial symbionts from this fish. Luciferin and luciferase have therefore been found only in beetles (both elaterids and lampyrids), molluscs (Pholas), ostracods (Cypridina, Pyrocypris), certain worms (Odontosyllis) and decapods (Acanthephyra). They cannot be demonstrated in bacteria, fungi, sponges, radiolaria, flagellates (Noctiluca), medusae, pennatulids, ctenophores, ophiurians, tomopterids, chaetopterids, terebellids, polynoeids, earthworms, schizopods, myriapods, cephalopods (*Watasenia* and *Heteroteuthis*), ascidians (*Pyrosoma*), balanoglossids or fish. The smaller dinoflagellates, springtails, flies, *Phyllirrhoé* and a few others have never been tested.

When *Cypridina* luciferin is mixed with extracts of these other luminous animals so prepared that they should contain luciferase, no light appears. There is also no light if *Cypridina* luciferase is mixed with extracts of the other forms so prepared that they should contain luciferin. Consequently the reactions are highly specific. Only if the luciferin and luciferase of closely related forms, such as two different genera of ostracods (*Cypridina* and *Pyrocypris*) or of fireflies (*Photinus* and *Photuris*) are mixed, will luminescence occur. No light appears if photogenic substances of the firefly are mixed with those of ostracods. *Cypridina* luciferin will not even give light with *Acanthephyra* luciferase (or vice versa), both crustacea, but belonging to different orders.

The firefly experiments are of particular interest since the genus *Photinus* has a reddish light, the genus *Photuris* a yellowish light. As the author (1917) has shown by "crossing" experiments, the animal supplying the luciferase determines the color of the resulting light, thus indicating that luciferase molecules actually emit the luminescence (see p. 155).

PROPERTIES OF PHOTOGENIC SUBSTANCES

It is obvious that the luciferins and luciferases from different animals are different and should be designated by the prefix, *Cypridina* luciferin, *Pholas* luciferase, etc. The properties of various kinds of photogenic substances are as follows:

Firefly luciferin and luciferase. Dubois did not study the chemical properties of luciferin and luciferase from *Pyropho-*

CHEMISTRY OF LUMINESCENCE

rus, the first form with which he worked, except to point out that *Pyrophorus* luciferase was destroyed on heating and was precipitated by alcohol while the *Pyrophorus* luciferin was not so affected. Luciferin was found only in the luminous organ of *Pyrophorus*, not in the blood; luciferase probably exists throughout the animal.[4]

McDermott (1915) and Harvey (1915, 1917) have attempted to make extracts of firefly material, but the results were mostly negative. Although dried firefly organs will luminesce on adding aerated water, it was not possible to obtain a water extract in absence of oxygen which would luminesce when aerated. Apparently the photogen decomposes. No luminescent material that would luminesce on adding water could be extracted from dry firefly organs by hot or cold ether, chloroform or alcohol or by cold acetone, toluol or carbon tetrachloride. Firefly luciferin is not readily oxidized on boiling and is found throughout the body of the firefly, while luciferase is restricted to luminous regions. The part played by the granules in light production needs further study and it is possible that what is regarded as luciferin in the firefly is really a photophelein.

Pholas luciferin. In a series of papers (1887-1918) Dubois has studied the chemical properties of *Pholas* luciferin and *Pholas* luciferase. He finds the luciferin to be destroyed above 70°C., to dialyze slowly, to oxidize with light production in the presence of *Pholas* luciferase, potassium permanganate, hydrogen peroxide, haematine and hydrogen peroxide, barium and lead peroxides, hypochlorites and the blood of various marine molluscs and crustacea.[5] It is insoluble in fat solvents,

---

[4] Private communication from R. Dubois.

[5] The author has been unable to confirm the presence of substances in blood of non-luminous forms or in non-luminous parts of *Pholas* that cause luminescence with *Pholas* luciferin. There are present the photophe-

but forms a colloidal solution in water from which it is precipitated unchanged by picric acid, alcohol, and ammonium sulphate. It is not precipitated by sodium chloride or acetic or carbonic acids, except in presence of neutral salts. It forms an insoluble "alkali albumin" with ammonium hydroxide. At various times Dubois has regarded luciferin as a proteose, as a nucleoprotein, and as a natural albumin having acid properties. It occurs only in the luminous parts of *Pholas*, not in non-luminous animals.

Pholas luciferase. Dubois finds that *Pholas* luciferase has all the properties of an enzyme, is destroyed at 60°C.; is non-dialyzable and insoluble in fat solvents, but forms a colloidal solution in water. It is not affected by 1 per cent sodium peroxide but its activity is suspended in saturated salt solutions, sugar or glycerine, and it may be preserved in this way, its activity returning on dilution. It is digested by trypsin and slowly destroyed by the fat solvent anesthetics, such as chloroform. Because he found iron in an extract of *Pholas* dialyzed for a long time against running water, Dubois considers that luciferase is associated with iron, and reports that it will oxidize the ordinary oxidase reagents, such as pyrogallol, gum guaiac, α-naphthol, paraphenylenediamine, etc. It remains to be proved, however, that luciferase, and not the oxidizing systems such as occur in all cells, is responsible for the coloration of these reagents. Dubois has found luciferases or substances capable of giving light with *Pholas* luciferin in the blood of many non-luminous crustacea and molluscs.[6]

Cypridina luciferin. Extracts containing only one part of dry whole *Cypridina* in 400,000,000 parts of water will luminesce sufficiently to be detected by the dark adapted eye. If the dried

leins, previously described, that cause luminescence in extracts of crude luciferase.

[6] See footnote 5, p. 135.

animal contains 1 per cent luciferin,[7] the concentration would be one in 40,000,000,000 (Harvey, 1923). These dilutions are much greater than those in which the color of intense dyes can be detected (1:10,000,000) and are comparable with dilutions in which the fluorescence of compounds like fluoresceine (1:10,000,000,000) can be seen. They are partly responsible for the difficulties encountered in manipulation and estimation of luminous extracts.

Chemical studies on *Cypridina* material have been made by Harvey (1917-1928), by Kanda (1919-1932) and especially by Anderson (1933-1937) who has succeeded in obtaining a highly purified extract. *Cypridina* luciferin is soluble in water, dilute salt, acid and alkali solutions; practically completely precipitated from a crude aqueous extract by phosphotungstic, flavianic acid, and tannic acid but not by picric acid; practically completely salted out by saturation with ammonium sulphate but not by magnesium sulphate or sodium chloride. It readily oxidizes spontaneously without luminescence in alkaline solution, less readily in acid solution and is stable for years[8] in water in absence of oxygen. *Cypridina* luciferin is dialyzable, readily adsorbed on surfaces and not destroyed by trypsin; it is soluble, as it occurs in the animal, in absolute methyl, ethyl, propyl and butyl alcohol and acetone but insoluble in ether, benzene, petroleum ether, chloroform, carbon disulphide, carbon tetrachloride, toluene and xylene.

Kanda has found that if fat-free dried *Cypridinae* are extracted with methyl alcohol and the methyl alcohol concentrated, some foreign material can be precipitated with ethyl alcohol. The filtrate contains luciferin and more foreign ma-

[7] Anderson (1935) finds about 0.15 per cent luciferin in dried *Cypridina* (see p. 138).

[8] A sample of luciferin, kept under a two-inch layer of vaseline, gave a good luminescence when mixed with luciferase after twenty-one years!

terial, which can now be precipitated with excess of benzene, leaving the purified luciferin in the filtrate. In a detailed study of solubilities, Anderson has confirmed the solubility of luciferin in benzene after preliminary extraction with methyl alcohol but finds that only part of the luciferin dissolves in the benzene with each precipitation and he could not purify the material more than 15- to 30-fold by this method.

The most successful method of preparation found by Anderson is to extract dried powdered *Cypridinae* with methyl alcohol in absence of oxygen. Ten per cent of butyl alcohol is added and the methyl alcohol removed in vacuo. The supernatant butyl alcohol extract is chilled and benzoylated with benzoyl chloride. After 15 minutes this solution is diluted with ten volumes of water and the new inactive luciferin derivatives extracted with pure ether. After removing the ether in vacuo the residual liquor is hydrolyzed with hydrochloric acid in absence of oxygen. The free active luciferin is left in the acid solution and can be extracted with butyl alcohol. By repeating the benzoylation and hydrolysis, a product purified 2,000-fold, as compared with dry *Cypridina*, can be obtained.

Luciferin is probably to be identified with the yellow granules of the gland cells of *Cypridina*. At least a yellow pigment must be concerned in luminescence for a marked yellow color is present in the luminous glands of many luminous animals—the worm, *Tomopteris*, in the medusae, *Aequorea* and *Mitrocoma*, and in copepods. The luminous organ of the firefly, of *Pyrophorus* and *Pyrosoma*, the slime of myriapods and of the fish, *Malacocephalus*, are yellowish. All attempts to demonstrate carotene pigments have failed but it is very probable that some flavin is present and possibly concerned in light production. It is unlikely that luciferin is a flavin but highly purified solutions are definitely yellow. Their absorption spectrum has been studied by Chase (1940), who found that an absorption band with a maximum at about $\lambda = 0.48\mu$ appeared in the first step of oxidation. During further oxidation, the absorption band disappears. The rate of color change paralleled the oxidation of luciferin in water at different H-ion concentrations, as well as in butyl alcohol as a solvent.

Since luciferin absorbs, it is not surprising to find that it undergoes non-luminous photochemical oxidation in strong light (1925, 1926). This is best seen by passing a beam of light from a carbon arc through the center of a test tube filled with the crude luminous extract. The region in the center quickly becomes dark. It can be shown that the darkening is due to more rapid spontaneous oxidation (without luminescence) of luciferin. The light does not affect luciferase. The inhibiting wave lengths are in the blue and near ultraviolet regions. Various dyes, such as eosin, cyanosin and methylene blue, will also act as photo-sensitizers to make the red, orange, yellow and green effective. Chase (1940) has recently found that the effect of light in accelerating oxidation of luciferin is dependent on the presence of a naturally occurring fluorescent pigment, probably a flavin. The light of a luminescent mixture of partially purified luciferin and luciferase was not quenched by strong illumination unless boiled *Cypridina* extract or pure riboflavin had been added.

The chemical nature of luciferin is as yet unknown but it is not a proteose, as the author at first (1919) supposed, and behaves like a compound of relatively low molecular weight. No antibodies for luciferin could be formed in the body of a rabbit (Harvey and Deitrick, 1930). Kanda (1932) who thought of luciferin as a phospholipin, reported crystallization but no details are given and no photographs of crystals have ever been published. Anderson (1936) believes that it may be a polyhydroxy-benzene, many of which have two step oxidations and redox potentials in the same region as that of the luciferin system.

Cypridina luciferase. This component is non-dialyzable; destroyed by trypsin; insoluble in alcohols and all fat solvents; soluble in water, in dilute salt, in acid and alkaline solutions, but active only over a certain P_h range; practically completely precipitated by phosphotungstic, tannic and picric acid and basic lead acetate; salted out by saturation with ammonium sulphate but not with sodium chloride; readily adsorbed on fine particles. It is probably to be identified with the small colorless granules of the luminous gland cells of *Cypridina*. In chemical deportment, luciferase behaves as a protein and an enzyme. The definite formation of an antiluciferase (not destroyed at 61°C. but thermolabile at 71°C.), when luciferase is injected into the blood of a rabbit, would indicate protein nature or very close association with a protein (Harvey and Deitrick, 1930).

The usual methods for purification of proteins can be used for luciferase. In most experiments it is sufficient to dialyze a cold, well stirred water extract of dried powdered Cypridinas against cold running water for 20 hours. A few drops of toluene added to this solution will preserve it for months with little loss in activity if kept in a refrigerator.

PROLUCIFERIN

In many enzyme reactions, the substrate, merely a link in a chain of reactions, is formed from some previous compound. It is obvious that luciferin must also be formed from some precursor in the cell and following the usual biochemical terminology, Dubois has called it *proluciférine* or *preluciférine*, and believed that it is converted into luciferin by another enzyme, coluciférase.[9] The experiments to prove the existence of proluciferin were first made by Dubois on *Pholas* in 1907 and have since been amplified (1917, 1918). Proluciferin is prepared from exhausted siphons or can be obtained on boiling an extract of the luminous organ of *Pholas*, because luciferin (at 70°), luciferase (at 60°) and coluciferase are all destroyed below the boiling point.

Coluciferase is prepared (1) by heating a *Pholas* luciferase solution to 65° (1917) or (2) by extracting with water, portions of the siphon of *Pholas* which have previously been macerated and well extracted with alcohol (1918). Long-continued treatment with alcohol apparently destroys the luciferase without affecting the coluciferase. On mixing a solution of proluciferin with one of coluciferase and allowing it to stand for 8 to 10 hours, luciferin is formed and can be recognized by the fact that it will give light with a crystal of potassium permanganate. Proluciferin does not luminesce with permanganate.

OXIDIZED LUCIFERIN

I believe that in the above experiments, Dubois was working with an oxidation product of luciferin, which I have called *oxyluciferin*, rather than a pro-substance. The mode of preparation both of *Pholas* proluciferin and *Pholas* coluciferase was such as could be used in the preparation of *Cypridina* oxyluciferin, and in a later paper Dubois (1919) took the view that his coluciferase was a reducing enzyme which formed luciferin by reduction.

It is, of course, obvious that when luciferin oxidizes, some oxidation products must be formed and it is certain that *Cypridina* luciferin oxidizes in two ways: (1) spontaneously,

---

[9] Dubois's use of the word coluciferase does not conform with modern usage. Gerretsen (1920) has called the enzyme that forms the photogen, photogenase, and believes the photogen is then oxidized by an oxidase.

without light production, and (2) in presence of luciferase, with luminescence. The author's attention was first directed (1918) to the possibility of reducing the oxidation product of luciferin by the observation that a test tube of clear solution of crude *Cypridina* luciferase (which must contain the products of oxidation of luciferin), although it may give off no light at first when shaken, after standing a day or so, emits a quite bright light on shaking. This was especially true when the luciferase had become turbid and ill-smelling from the growth of bacteria. Thinking that the bacteria produced a substance which could be oxidized by the luciferase, the author tried growing bacteria and also yeast on appropriate culture media, and, after some days of growth, mixing the culture media containing the products of bacterial or yeast growth with luciferase, expecting to obtain light; but no light appeared. However, if a little crude luciferase solution was added to the bacterial or yeast cultures and they were then allowed to stand a short time, light appeared whenever they were shaken. Indeed, such cultures behaved much as a suspension of luminous bacteria which has used up all the oxygen in the culture fluid and will only luminesce when, by shaking, more oxygen is dissolved. The effect turned out to be due to the anaerobic conditions, for the reducing action of bacteria and yeast is well known. Oxidized luciferin in the luciferase solution was reduced to luciferin. Minced muscle can take the place of yeast, for it also has a strong reducing action.

It soon developed that many methods of reduction which involved the addition of hydrogen could be used—finely divided platinum or palladium in presence of hydrogen, atomic hydrogen evolved from dissolving metals, such as magnesium powder or zinc dust. Indeed, if a freshly cut surface of aluminium, magnesium, manganese, zinc or cadmium is placed in a solution containing oxyluciferin and luciferase, the surface will luminesce. This is the result of reduction of oxyluciferin by a layer of nascent hydrogen near the surface of the metal, and reoxidation with light production in a more external layer. For the same reason, light appears at the cathode when a solution of oxyluciferin and luciferase is electrolyzed. Sodium hydrosulphite ($Na_2S_2O_4$), hydrogen sulphide, ammonium sulphide and other reducing agents are also able to reform luciferin from its oxidation product.

Since all methods of reducing oxyluciferin are similar to those by which dyes like methylene blue can be reduced, I compared the oxidation of luciferin to the oxidation of a leuco-dye, particularly methylene white to methylene blue. The oxidation of this dye takes place spontaneously in presence of molecular oxygen and more rapidly in alkaline solution. Chase (1940) has recently found that rates are almost identical, for both luciferin and methylene white oxidation are roughly proportional to the fifth root of the OH-ion concentration.

The oxidation of both is also affected by light (Harvey, 1925, 1926) and neither the oxyluciferin nor the oxidized methylene blue can be reduced by mere removal of oxygen in solution with an air pump. The change on oxidation is therefore not similar to that of hemoglobin. It is not an oxygenation:

Hemoglobin (Hb) + O_2 = Oxyhaemoglobin (HbO_2)

but may be represented as a dehydrogenation:

Luciferin (LH_2) + $1/2\ O_2$ =
    Oxidized luciferin (L) + H_2O
Leuco-methylene blue (MH_2) + $1/2\ O_2$ =
    Methylene blue (M) + H_2O.

Dehydrogenation changes are fundamentally due to electron transfers.

LIVING LIGHT

Reductant = Oxidant + H_2 or
Reductant = Oxidant + 2e.

The intensity of these systems on the oxidation-reduction scale is expressed by a redox potential. Methylene blue under standard conditions ($P_h = 7$, and equal concentration of oxidant and reductant) has an $E_o' = +0.025$ volt.

It early became clear (1927) that the luciferin system did not behave like the methylene blue system, in that it was not completely reversible and could be designated only by "apparent" oxidation and reduction potentials.

Anderson's (1936) work on purified luciferin has indicated two important facts regarding the oxidation. First, that the oxidation with light production in presence of luciferase gives an oxidation product which cannot be reduced. It is the oxidation without light production by oxidants like potassium ferricyanide which is reversible. Second, that the change with ferricyanide occurs in two steps, one the reversible oxidation previously referred to, the second irreversible, and probably also an oxidation, although this has not been definitely demonstrated. The spontaneous oxidation of luciferin without emission of light in crude solutions (without luciferase) is probably catalyzed by traces of heavy metals in the solution and proceeds much more slowly when the luciferin has been purified. Both the dark oxidation and the luminescent oxidation must take place simultaneously when luciferin is mixed with luciferase.

The redox potential of the first step has been placed by Anderson (1936) near the hydroquinone ⇌ quinone system but 0.01 ± 0.005 volts more negative. Korr (1936) also considers it near quinone. If we assume a two electron change, the redox potential would be $E_o' = +0.26$ at $P_h = 7.0$, about halfway between quinone and orthochlorphenol-indophenol.

CHEMISTRY OF LUMINESCENCE

The reversible product has been designated by Anderson as oxidized luciferin, a name indicating that it is the oxidant (while luciferin is the reductant), since oxyluciferin implies that the change is an oxygenation, like oxyhemoglobin. The nature of the oxidant formed during the luminescent reaction with luciferase is unknown.

A large number of my early experiments (see 1928) were directed to finding a means of oxidizing Cypridina luciferin with light production in absence of Cypridina luciferase. These included extracts of luminous and also non-luminous forms, extracts which must have contained the iron respiration catalyst or the yellow respiration enzyme; plant extracts (potato and turnip juice) containing oxidases, peroxidases and catalases, tried with and without H_2O_2; hemoglobin, with and without H_2O_2; a long list of oxidizing agents from all positions on the redox scale, permanganates, bichromates, ferricyanides, barium peroxide, benzoyl peroxide, perchlorates, persulphates, perborates, hypochlorites, hypobromites, hypoiodites and various heavy metal catalysts such as iron, copper and manganese. Oxidation of luciferin occurs but no light is emitted.

A similar attempt to find easily oxidizable substances that will luminesce with Cypridina luciferase also failed. Included were compounds known to exhibit chemiluminescence, such as various oils, esculin, lophin, amarin, aminophthalichydrazid (luminol), dimethyldiacridylium nitrate, pyrogallol and many easily oxidizable amino- and hydroxy-phenol compounds; also many fluorescent dyes. The same specificity that prevents luciferin of one form from luminescing with the luciferase of another probably prevents these easily oxidizable compounds from giving light.

LUCIFERESCEINE

In Pyrophorus and fireflies, as in many non-luminous insects, there occur fluorescent substances. Dubois (1886) called the material from *Pyrophorus, pyrophorine,* and believed that it intensified the light and changed its color by transforming invisible into visible rays. Coblentz (1909) and McDermott (1911) found a similar compound which can be extracted from fireflies and called it *luciferesceine,* by analogy with fluoresceine. Its emission spectrum is complementary to that of the firefly luminescence (Ives and Coblentz, 1910), in the blue region, so that, according to Stokes's law, it could not be excited by the longer wave lengths in the chemiluminescent spectrum of the firefly. Luciferesceine is found abundantly in non-luminous parts of the firefly and probably plays no part in luminescence, despite its name.

It is to be expected that a substance excited to luminesce by the energy of a chemical reaction would also show fluorescence, excitation by light itself. This is true of many chemiluminescent compounds or their oxidation products. If the firefly or *Pyrophorus* is examined in ultraviolet light from which visible wave lengths have been removed by a nickel oxide ultraviolet filter, the luminous organ fluoresces with an intensity and color that closely approaches the chemiluminescence of these forms. It is very possible that an oxidation product of firefly luciferin is the fluorescent material, for in ctenophores, fluorescence only appears after the animal has been stimulated to luminesce. The fluorescent material disappears in 10 to 15 minutes and upon a second stimulation, fluorescence again appears (Harvey, 1925). Other forms, whose luminous secretion is fluorescent, are ophiurians, the worm, *Acholoë,* and copepods. Many deep-sea fish (even those preserved in formalin) have brightly fluorescent luminous organs, as have some squid; but luminous bacteria, fungi, pennatulids, the medusa, *Pelagia,* worms such as *Thelepus* and *Tomopteris, Pholas* and *Cypridina* are not markedly fluorescent in ultraviolet light, either before stimulation or after an oxidation product has been formed (Harvey, 1926).

MODELS OF BIOLUMINESCENCE

From the foregoing account it will be seen that light production in different animals differs in detail but that a general plan is followed. As Dubois (1914) expressed it, we are dealing in luminous organisms with "1° une luminescence; 2° une chimieluminescence; 3° une oxyluminescence; 4° une zymoluminescence. Ou si l'on veut bien admettre que les zymases sont encore quelque chose de vivant, une Biozymoöxyluminescence." Perhaps it is no longer necessary to admit that the enzymes are living in order that we may adequately visualize the nature of the photogenic process.

We have already seen that many organic compounds are capable of a bright luminescence when oxidized under proper conditions with various oxidizing agents. Some require strong oxidizing agents that would have a very deleterious effect on living cells but others can be oxidized in a manner that exactly imitates the light production by animals.

In 1913, Ville and Derrien, in a short note to the French Academy, "Catalyse Biochemique d'une Oxydation Luminescente," showed that lophin could be oxidized with luminescence by vertebrate blood in the presence of hydrogen peroxide. In the same year Dubois (1913) found that esculin, the glucoside from horse-chestnut bark, would also oxidize and luminesce in presence of blood and hydrogen peroxide. In these cases the hemoglobin of the blood has peroxidase action, transferring oxygen from the hydrogen peroxide to esculin or lophin, and is thus comparable to luciferase, except that luciferase does not require the presence of hydrogen peroxide.

As the hemoglobin does not lose this power on boiling, whereas luciferase does, the analogy is not perfect. Many oxygen catalysts are known, however, which may be destroyed on boiling, namely, the peroxidases of plant juices. Esculin will not luminesce with peroxidase and hydrogen peroxide, but pyrogallol or gallic acid will (1916). If one mixes dilute pyrogallol solution + hydrogen peroxide with potato or turnip juice or almost any plant extract, a yellowish luminescence appears. The plant extract loses the power to cause such luminescence on boiling and the peroxidase will not dialyze. It is, of course, comparable to luciferase and acts on the thermostable dialyzable pyrogallol–hydrogen peroxide mixture, which is comparable to luciferin. Luminol can be substituted for pyrogallol in the above experiments.

KINETICS OF LUMINESCENCE

In the study of kinetics of ordinary enzyme reactions, the concentration of the substrate must be determined at definite intervals of time. Such a procedure is not possible with luciferin, as methods of analysis, apart from the luminescence it gives, are unknown. However, when luciferin and luciferase solutions are mixed, we observe a bright luminescence which represents light due to some initial concentration of luciferin. After mixing, the luminescence gradually decreases in intensity and a study of this decay curve, both with respect to form and area involved, has given an interesting insight into the course of the oxidation. Quantitative studies on photogenic substances therefore require measurements of luminescence intensity (rate) and total luminescence. They are unique because they involve a direct measurement of reaction velocity apart from any determination of the concentration of reacting substances.

CHEMISTRY OF LUMINESCENCE

Not only is photometry of relatively weak lights uncertain, but becomes much more so if the light intensity is continually changing, continually decreasing, as it is during decay of Cypridina luminescence. The problem is a difficult one, first solved by Amberson (1922) by a photographic method. Amberson photographed the light on a moving picture film wrapped on the drum of an ordinary kymograph. The container for the luminescent mixture was a vessel of glass, covered with black paint and adjusted very near the drum. On one side was a small slit, through which the light shone directly on the film. A stirring device and thermometer were placed in the tube and a pipette containing the luciferase fixed in such a position that luciferase could be mixed with luciferin in the vessel at the proper time. Records taken in this way during revolution of the drum showed, on development, a streak of blackening, which gradually faded out during revolution of the film, as the luminescence became less intense. Calibration exposures had to be placed on the same film in order to determine how much blackening a given intensity of light would produce. They were obtained by allowing Cypridina luminescence to pass through neutral filters of known absorption before striking the film. Densities on the film were then read as relative light intensities by an optical pyrometer or some other device. Oxygen was always present in such high concentration that it did not affect the light intensity even though its concentration varied somewhat during the course of the reaction.

Curves obtained in this way can be described by the equation for a unimolecular reaction. They are straight lines if log intensity of luminescence is plotted against time (Fig. 76). Since this decay curve is the same as would be followed when reaction velocity, dx/dt, is plotted against time, the conclusion can be drawn that intensity of luminescence (with-

LIVING LIGHT

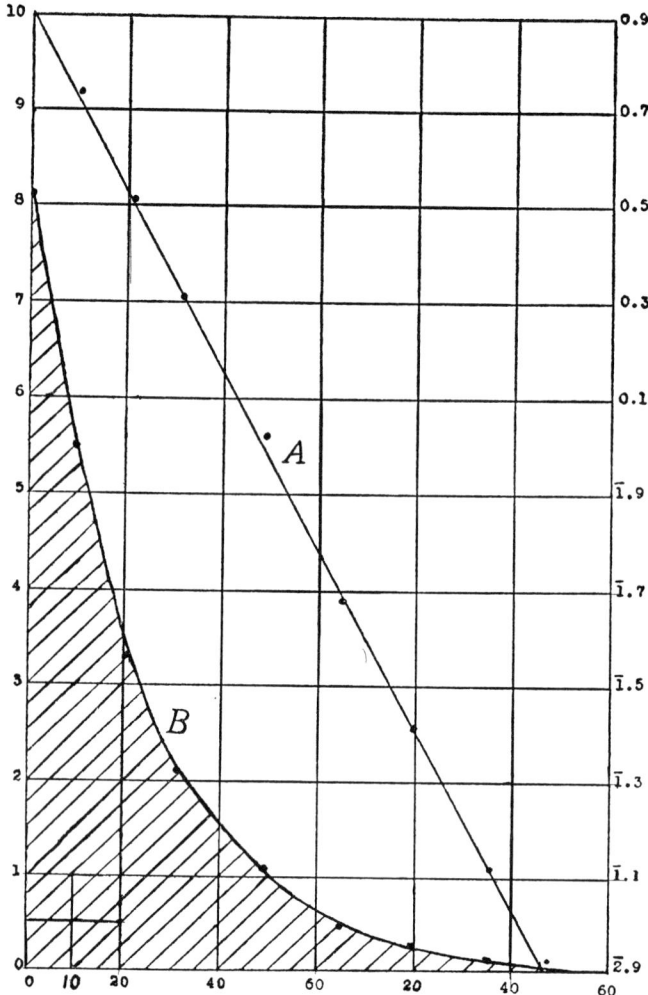

FIG. 76 The decay curve of luminescence, B, when *Cypridina* luciferin is mixed with small amounts of *Cypridina* luciferase. Luminescence intensity vertical; time in seconds horizontal. The reaction is half complete in 17 seconds. The area of the curve measures the total light emitted. Note that plotting log luminescence intensity against time, A, gives a straight line. After Stevens.

CHEMISTRY OF LUMINESCENCE

in certain limits) is determined by rate of oxidation of luciferin, and that only one molecule limits the reaction. Plotting log concentration of luciferin against time should also give a straight line for a unimolecular reaction. However, the experiments in which luciferase concentration or temperature is varied while initial luciferin is the same, show that luminescence intensity is determined by reaction velocity, dx/dt, and not by concentration of luciferin.

The principal facts established by Amberson's work are:

1. Stirring does not affect the luminescence intensity or the form of the decay curve.

2. The slope of the decay curve is proportional to luciferase concentration, i.e. the velocity constant is proportional to luciferase concentration.

3. Logarithmic plottings with two different concentrations of luciferin (made by allowing the luciferin to oxidize spontaneously to different degrees), but with the same luciferase are parallel. The velocity constant is not affected, but light intensity (reaction velocity) is naturally less with smaller concentrations of luciferin. If the luciferin solutions are diluted with oxidized luciferin solution, or with water, the velocity constants become greater with increasing dilution.

4. There is an initial flash of light when the luciferin and luciferase are mixed, too high to agree with the remainder of the curve. Its meaning is unknown.

5. The temperature coefficient is high. Values obtained in nine experiments give an average Q_{10} of 2.74 for different 10° temperature intervals. There is a tendency for the higher (25°-35°) intervals to have a greater Q_{10} than the lower (5°-15°).

Making use of a direct photometric method, Stevens (1927) obtained similar kinetic results. He studied mixtures of luciferin and luciferase whose half decay time was about 24 seconds, three times as long as in Amberson's experiments.

If a large amount of luciferase is added to luciferin, the time for half decay may be very short, 0.5 to 1.0 second. A photocell method of recording is necessary, with amplification of the small currents and a short period galvanometer whose deflection is registered by a light-beam marking on moving film or photographic paper. Harvey and Snell (1930) used a photocell-amplifier-string galvanometer combination to record these rapid flashes of luminescence whose decay is reproduced in Fig. 77. Again the diminution in luminescence was logarithmic (except for large ratios of luciferin to luciferase) and the velocity constant of rapid flashes of luminescence was proportional to enzyme concentration. This constant varied approximately inversely as the square root of the total luciferin (luciferin + oxyluciferin) concentration.

The total light emitted is an important quantity. Theoretically, it should be a measure of the initial quantity of luciferin, provided the reaction is unimolecular and luminescence intensity is a measure of reaction velocity, as is the case. Practically, total light can be obtained from graphical integration of the area under any decay curve, but this is a laborious calculation. Anderson (1933) has perfected a simple method of directly measuring total amount of light emitted under various conditions, for routine quantitative determination of luciferin and luciferase. The total luminescence is detected by a photocell, and the photoelectric current allowed to charge a condenser to a voltage which is read on a potentiometer, using the Lindemann electrometer as a null instrument. Condenser voltage is a measure of luciferin. Rate of change of condenser voltage gives the velocity constant of the reaction and hence the luciferase concentration. Moreover, Anderson used greatly

CHEMISTRY OF LUMINESCENCE

purified luciferin and luciferase, an important advance over previous work. His results were as follows:

(1) The total light emitted under uniform conditions was approximately proportional to the amount of luciferin initially present, and independent of the concentration of luciferase.

(2) From 18° to 28° C. the total light emitted per unit of luciferin decreased from 2 to 3.5 per cent per degree increase in temperature.

(3) The total light varied with H-ion concentration. It was less at $P_h = 7.8$ than at $P_h = 6.8$ and less at $P_h = 6.8$ than at $P_h = 6$.

(4) The total light was dependent on the salt content of the medium. Less was emitted in m/60 than m/15 phosphate buffer at the same P_h and less from m/20 phthalate buffer than from m/20 phosphate buffer at $P_h = 6$. More total light was emitted in presence of 0.34 m NaCl than in 0.02 m NaCl. A more recent research (Anderson, 1937) has dealt with these salt effects in detail. Kinetics have also been studied in heavy water (Anderson and Harvey, 1934).

Although solvent conditions modify the luminescence values, under definite standard conditions of temperature, P_h and salt content of medium, total light may be used as a quantitative measure of luciferin and the velocity constant as a quantitative measure of luciferase concentration. The kinetics of luminescence is so greatly affected by the presence of added salt and by minute amounts of heavy metals and other impurities that the greatest care must be taken in any research of this kind. Some anomalous delayed and long lasting luminescences have been explained by Anderson (1936) as due to reduction of oxidized luciferin, which is then slowly reoxidized with emission of light.

None of the above kinetic studies measures the rate of appearance of light, since in the ordinary methods of mixing luciferin and luciferase this velocity is more rapid than the time of mixing. By a special technique developed by B. Chance, the author together with Johnson, Millikan and Chance (1940) have determined the time for appearance of light of very rapid flashes. The time to reach half maximum intensity is 0.006 second and full intensity 0.03 second. The decay in these ultrashort flashes, which fall to half intensity in 0.12 second, is also logarithmic.

The above statement applies to conditions where oxygen is already present. If luciferin and luciferase are previously mixed in absence of oxygen and then this anaerobic solution mixed with aerated water, the time to half maximum intensity is 0.002 second and to full intensity 0.008 second, three times as fast. The experiment clearly shows that luciferin and luciferase combine slowly, whereas the combination with oxygen is rapid.

It is interesting to compare this reaction rate with the development of light when luminous bacteria in absence of oxygen are suddenly mixed with aerated sea water. The time to half maximum is 0.08 second and to peak intensity 0.34 second, very much slower than the corresponding process for *Cypridina* luminescence. As oxygen tensions varying between 380 mm. Hg and 10 mm. Hg did not alter the time relation, it would appear as if the slow rate of appearance of luminescence in bacteria were a characteristic of the oxidation in these forms and not due to time of diffusion of oxygen through the cell surfaces.

MECHANISM OF LUMINESCENCE

The above studies on kinetics supply good evidence that chemiluminescence intensity is determined by a rate of reaction. The view is not new, for Trautz (1905), came to this conclusion from his extensive study of the chemiluminescence

CHEMISTRY OF LUMINESCENCE

of phenol and aldehyde compounds. He based his views largely on the effects of temperature and concentration of reacting substances and went so far as to declare that any reaction would produce luminescence if the reaction velocity were sufficiently increased. This statement appears to be too extreme. It is a *particular* reaction occurring at an optimum velocity that results in light. We now believe that such a reaction leads to excited molecules having excess energy. The reaction products may themselves become excited and emit radiation on returning to the normal state (direct chemiluminescence), or they may, by collision, hand over the excess energy to other molecules (sensitized chemiluminescence).

These principles are well illustrated by *Cypridina* luminescence. Luciferin can be oxidized rapidly (at high temperatures or by ferricyanide) without light; or it can be oxidized slowly with light emission, provided luciferase is present. Only then is light intensity determined by a rate of reaction and proportional to the concentration of luciferase. Luciferase evidently plays a double rôle—both that of oxidative catalyst and that of supplying molecules which can be excited to luminescence. This may be represented by the following scheme, which expresses, as far as is possible, the mechanism of the luminescent reaction. Luciferase is a catalyst for the first reaction.

Luciferin $(LH_2) + 1/2\ O_2 \rightleftarrows$ excited oxidized luciferin $(L') + H_2O$

L' + luciferase $(A) = L$ + excited luciferase (A')

$A' = A + h\nu$

The picture in *Cypridina* is perfectly definite, but in the majority of luminous animals, especially forms like pennatulids, in which light emission is associated with granule breakdown, the luciferin-luciferase reaction cannot be demonstrated. Assuming that luciferin and luciferase exist, inability to prepare them might be due to (1) small amounts of these substances in the luminous cells (2) instability of one or both (3) just enough luciferase to react completely with luciferin, i.e. luciferase does not behave as a catalyst. In the last case it would never be possible to prepare a luciferase solution. Without going into the details of experimentation it seems most likely that the third possibility is the correct one and that the granules of pennatulids contain just enough luciferin and luciferase, separated from each other by a film, for complete reaction. On dissolution of the granule, the film breaks down, light appears, and no luciferase is left over. Even in *Cypridina*, luciferase will not oxidize an indefinite quantity of luciferin although a large excess is present. In forms where luciferin and luciferase cannot be demonstrated, we may suppose that the primary rôle of luciferase is not to act as a catalyst but to supply easily excited molecules.

In luminous bacteria, luciferase may exist as an endoenzyme, not extractable by ordinary methods, thus resembling most bacterial dehydrogenases. It has not been possible to obtain luminous extracts despite many attempts (Harvey, 1915; Korr, 1935). Every method that breaks up the bacterial cell permanently destroys its ability to luminesce. Quickly dried bacteria will give a temporary flash of light when moistened and some of these will grow on culture media but if the dried bacteria are first ground with sand, luminescence is weak or lacking when moistened.

The luminescence is not a *necessary* concomitant of the respiration of the bacteria, since non-luminous strains can be obtained by heat treatment (Beijerinck, 1912, 1915), but normally luminescence is linked with the general respiration. Some reagents (cyanide) markedly affect respiration without a large effect on luminescence, while others (lipoid soluble narcotics) markedly affect luminescence without reducing the

CHEMISTRY OF LUMINESCENCE

respiration. Further consideration of their effects will be given in Chapter V.

A very small per cent of the total oxygen consumption is used in luminescence but luciferin itself as a hydrogen donor may take part in respiration of foodstuffs by the bacterial cell. According to a scheme of van Schouwenburg (1938), some oxygen goes to oxidize foodstuffs through the ferment-haemin system (haemin respiration), some through a cyanide-insensitive respiratory ferment (rest respiration), and some through luciferin as hydrogen carrier ("light" respiration). The "rest" respiration is about 10 per cent of the total at all oxygen tensions, but the per cent of haemin respiration decreases and the per cent of "light" respiration increases with increase in oxygen tension. Basing his theory on the two types of oxidation of Cypridina luciferin, van Schouwenburg believes that the "light" respiration is made up in part of luciferin, oxidized reversibly by oxygen without light production and in part oxidized in presence of luciferase with luminescence.

Exhaustive studies by Johnson (1939) and by Johnson, van Schouwenburg and van der Burg (1939), in which the total light and the excess light which appears after lack of oxygen are recorded, have led to the following scheme for bacterial luminescence:

(1) $L + XH_2 \rightleftarrows LH_2 + X$
(2) $LH_2 + A \rightleftarrows A.LH_2$
(3) $A.LH_2 + 1/2\ O_2 \rightarrow A + L_1 + H_2O + h\nu$

The symbol XH_2 is some hydrogen donor; LH_2, luciferin; L, the oxidation product without light production; L_1, the oxidation product formed when luminescence occurs; and $A.LH_2$, a combination of luciferase (A) and luciferin necessary for luminescence. Like all attempts to visualize mechanism, it is a working hypothesis which may stand or fall with the results of future investigation.

EVOLUTION OF LIVING LIGHT

The widespread but "spotty" distribution of luminescence in the living world was emphasized in Chapter II. We have just seen what chemical mechanisms are necessary for the emission of light and may now inquire how the evolution of such a remarkable property is to be visualized. Perhaps the following, which the author suggested in 1932, will serve as an outline of this fascinating problem.

"In the simplest organisms, the bacteria and the fungi, light production evolved in connection with the cell respiratory mechanism by the development of one of the hydrogen acceptors whose oxidation gives sufficient energy to excite a compound similar to luciferase, possibly some special protein within the bacterial cell. Chemiluminescence studies have shown that fluorescent compounds, i.e. substances which can be excited to luminescence by the energy of radiation, are also most likely to exhibit chemiluminescence. They are excited by the energy of a chemical reaction. Practically all proteins are fluorescent and all luminous cells are somewhat fluorescent. In addition, many luminous animals have unusually bright fluorescent compounds in their luminous organs. It is not, therefore surprising, it is likely that an unusually bright 'photophore' group in a protein should be seized upon and developed during the evolutionary process to the chemiluminescent substance, luciferase. Since the bacteria and fungi produce a steady luminescence which is unaffected by stimulation, this luminescence may be regarded merely as an accompaniment of the hydrogen acceptor mechanism; hence fortuitous and of no special use to the organism.

CHEMISTRY OF LUMINESCENCE

"The next step is the development of a mechanism for controlling the luminescence, for protozoa (dinoflagellates, cystoflagellates and radiolaria) have developed within the cells a means of glowing only on stimulation, which is comparable to the stimulating mechanism of a muscle or nerve cell. . . .

"Practically all higher forms also luminesce only on stimulation and various means have been devised to bring this about. In multicellular animals we find developed, special glands for secreting luminous material or special organs of luminescence with accessory structures; shutters, reflectors, lenses, pigment screens and color screens, the whole forming a lantern. Here we may see a use for luminescence, warning or frightening predaceous forms, luring food, illumination or recognition signals in the bringing together of the sexes.

"There has never been any great evolution in the direction of an entirely luminous group. Perhaps the coelenterates[10] most closely approach this, for luminescence is more widespread there than in any other phylum. Generally, luminescence has appeared sporadically in the living world, with a few luminous species scattered here and there among structurally very close non-luminous relatives. This again means that luminescence can be readily developed in the course of evolution by some slight change in a mechanism *already existing* within all cells. I believe this mechanism is the respiratory one and that luminescence has resulted from the transformation of some of the hydrogen acceptors of the cell, together with development of proteins with very actively fluorescent groups which are excited to luminescence by the energy of the oxidative dehydrogenation."

[10] Including ctenophores.

CHAPTER V

PHYSIOLOGY OF LIGHT PRODUCTION

STIMULATION

ABILITY to control their light is a fundamental characteristic of animals. So universal is this power that a steady luminescence in an animal can be attributed *a priori* to the presence of symbiotic bacteria. The only possible exception is the beetle, *Phengodes*, whose light is continuous, although it has not been studied sufficiently to state that it contains luminous bacteria (p. 70). Even when bacteria are present, we have seen that a secondary mechanism of controlling the light is often developed, as in the East Indian fish— *Photoblepharon*, by a movable screen, *Anomalops* by rotating its light organ, both nerve controlled.

Stimulation to light is to be compared with stimulation of a muscle to contract. The connection is best seen in *Noctiluca*, as was pointed out by Quatrefages (1850), and by Watase (1898) in a paper entitled "Protoplasmic contractility and phosphorescence." Nerve-muscle physiology has undergone a revolution in recent years. We now know fairly well the steps by which the store of chemical energy in a muscle is partially transformed into the mechanical energy of a movement. Two mechanisms are involved, that of shortening and that of stimulation.

Light production involves both processes also, the chemical reactions which emit light and the method of starting these reactions. The former were considered in Chapter IV and the latter will now be described. Stimulation to luminescence is a far more varied process than the stimulation of a muscle, for

PHYSIOLOGY OF LUMINESCENCE

different methods have been developed by luminous animals to turn their light on and off:

(1) If the luminous organ is a large gland whose cells are filled with luminous granules or if there is a reservoir, the secretion may be squeezed out by muscles, as in *Cypridina*. Possibly the deep-sea shrimp, *Acanthephyra*, some fish and the squid, *Heteroteuthis*, belong in this category also. No special problem of stimulation to luminesce is involved here, since the nerves do not directly end in photogenic or secretory cells.

(2) If the luminous material is formed in gland cells over the surface of the body, these cells may be stimulated to secrete by nerves. The problem is similar to that of secretion in general—the innervation of gland cells. *Chaetopterus* and many other worms, *Pholas*, myriapods, probably medusae and pennatulids and possibly ctenophores belong in this category.

(3) If the animals are characterized by intracellular luminescence, there may be nerve or hormone control of the organ. Forms with photophores belong here, many shrimp, squid and fish; also some insects such as fireflies and the elaterid beetle, *Pyrophorus*, possibly also the fly larva, *Boletophila*. The mechanism of flashing in the firefly has aroused much interest and various theories have been proposed which will be considered shortly.

(4) If the animals are protozoa, such as flagellates and the radiolarians, any kind of stimulation causes a flash of light which can be seen to come from minute granules scattered throughout the cell protoplasm.

In the first two groups, luminous material is merely poured outside the cell, where (1) luciferin and luciferase come in contact with each other or (2) photogenic granules break down with light emission. In the last two groups the problem is essentially how stimulation (direct or through nerves) starts a series of reactions, one of which emits light.

LIVING LIGHT

Since light production is dependent on oxygen, we can imagine (1) that stimulation admits oxygen to the cell by increasing cell permeability, or (2) that stimulation removes some phase-boundary between luminous substances so that they come into contact with each other. It can be shown (Harvey, 1922) that most cells are so readily permeable to oxygen that one cannot distinguish differences in permeability, even between dead and living cells. Therefore permeability for oxygen would appear to be independent of stimulation. If the luminescence of a cell is dependent on increase in oxygen supply by stimulation, the mechanism for increasing oxygen must be an external secondary one and not an increase in permeability of the photogenic cell membrane itself. In the firefly there could be such a device for regulating the oxygen supply to the photogenic cells by opening the ends of the air tubes but no mechanism of this kind is present in the free-swimming luminous protozoa like *Noctiluca*. The most suggestive line of inquiry points to some change in phase-boundaries in the cell bringing the substances concerned in light production in contact with each other.

Noctiluca is an ideal form for study of stimulation and may be taken as an example of the behavior of luminous animals generally. Quatrefages (1850), Vignal (1878), Massart (1893), Pratje (1921) have contributed to our knowledge while E. B. Harvey (1917) has made an extensive study.

Stimulation may be by mechanical, osmotic, chemical, thermal and electrical means. The first is most natural, agitation due to the breaking of waves, the last best suited to experimental analysis, since a comparison can be made with known current effects on other cells. To quote from E. B. Harvey:

"When a constant current is passed through a mass of noctilucas, the animals flash brightly at the make, continue glow-

PHYSIOLOGY OF LUMINESCENCE

ing during the passage of the current, and cease to glow at the break, giving no flash, but sometimes they stay glowing after the break, and in this case the stronger the current the longer the glow lasts. If stimulated mechanically while the current is passing, they respond by a flash, just as when no current is passing.

"The light comes from all parts of the *Noctiluca* and is not restricted to anode or cathode regions. No increase in luminosity could be observed on the cathode side nor decrease on the anode side of the animal comparable with the polar effects of the current on muscle."

There is thus a difference in behavior as compared with normal skeletal muscle which contracts only on make and break of a constant current, remaining at rest during the passage of the current. Sometimes, under special conditions, a frog's sartorius muscle will contract on the make, stay contracted while the current is passing and relax on the break. The tentacle of *Noctiluca* behaves in a similar manner.

"When a mass of noctilucas is subjected to the induced currents of an induction coil they respond with a flash at the break, and at the make also, if the current is strong enough.

"If subjected to an interrupted induced current for forty-five seconds, the animals flash on the first shock and then remain glowing, but the luminosity becomes gradually fainter. If the current is now stopped for a moment and then passed again, there is again a bright glow. The animals therefore fatigue rather readily when stimulated electrically, as they also do with mechanical stimulation.

"If a number of noctilucas are punctured with a needle, causing the cells to collapse, and are then subjected to an interrupted current, they respond just as uninjured cells do. Such punctured and collapsed cells likewise give a normal response when stimulated mechanically—i.e. by pressure of a cover-slip.

... It is thus shown that injury to a *Noctiluca* does not interfere with its response to mechanical or electrical stimulation.

"If, however, the injury to the cells is too great, and the cells are completely broken to pieces, they do not respond to stimulation. By pressing a mass of noctilucas through cheesecloth, a filtrate was obtained containing many empty membranes and fragments of cells, visible under the microscope. This filtrate, although luminous, did not respond to electrical stimulation. . . ."

In multicellular forms, such as coelenterates, stimulation may be direct or indirect; just as in a nerve-muscle preparation, contraction may result from direct stimulation of the muscle or indirect stimulation through its nerve. We have seen that stimulation of a nerve causes a wave of light to pass over the nerve net of these forms in any direction (Parker, 1920) as described in Chapter II. The speed of the wave and other phenomena are no different from the speed and properties of other nerve impulses. In *Renilla* the speed of a nerve impulse to muscles retracting the zooids is the same as that which stimulates to luminescence, 7.5 cm. per second. Possibly the same nerve is involved although mechanical stimulation readily results in luminescence, rarely in retraction of zooids. In some higher forms true photogenic nerves, nerves that supply light organs alone, are present.

Ctenophores are especially sensitive to mechanical stimulation but electrical stimulation also causes a wave of light to pass in either direction along the canals, where the luminous cells are situated. Peters (1905) found no correlation of luminescence with activity of the aboral sense organ. Moore (1924) also showed that in *Mnemiopsis*, tactile receptors for luminescence lie only along the rows of paddle plates, whereas tactile receptors for ciliary and muscular movement are distributed generally over the surface and connected by a nerve net.

PHYSIOLOGY OF LUMINESCENCE

In worms the distribution of definite photogenic nerves is more restricted.

In those shrimp, squid and fish which have a pattern of photophores, the control of light may be compared with the control of color patterns in animals. Luminous sharks have not been properly investigated but the enormous number of minute photophores bears a striking resemblance to chromatophores. Color changes due to expansion or contraction of chromatophores may be either nerve controlled or hormone controlled. Both methods may be involved in one species of fish. It is then not surprising to find that the luminescence of fish may be both nerve and hormone controlled. The first observations were those of Greene and Greene (1924) on the California toad-fish, *Porichthys*, extended by the author (1931) to the deep-sea fish, *Echiostoma tanneri*, already described in Chapter II. We have seen that adrenalin is able to excite all the organs of these two fish, a condition not entirely accomplished by the most violent mechanical or chemical stimulation. It would appear as if full luminescence were a reaction brought into play only in times of greatest excitement and had to do with escape from attack or with sexual recognition. Greene found that insulin injected into toad-fish did not stimulate to luminescence.

In the firefly the light is undoubtedly a signalling system between the sexes. Like incandescent signal lamps the light must go on and off rapidly. In the lamp this is accomplished by a thin ribbon filament that gains maximal heat rapidly, in 0.08 second.[1] The firefly does a little better. Photocell records of the flash of *Photuris pennsylvanica* show a rise to maximum light intensity in 0.06-0.07 second and a similar fall to zero light, the

[1] Data kindly supplied me by Dr. W. E. Forsythe of the Nela Research Laboratory, Cleveland, Ohio, from measurements of Mr. Hinman.

whole flash lasting 0.12-0.14 second, as in Fig. 78 (Snell, 1931-1932; Brown and King, 1931).

Almost every student of luminescence has been intrigued with the firefly, so that a mass of disconnected physiological facts have been accumulated by such workers as Spallanzani (1796), Carradori (1796-1799), Macaire (1821, 1822), Todd (1824, 1826), Kolliker (1858-1859), Heinemann (1872-1873, 1886), Jousset de Bellesme (1880), Dubois (1886) and Verworn (1892). Only the recent work can be considered in detail.

Flashing is reflex, from nerve stimuli whose receptor organs are visual or tactile. Flashing starts when the daylight intensity falls to a certain value at dusk or in the afternoon if storm clouds obscure the sun (Buck, 1937). Its rate depends on the temperature (Snyder, 1920). Blowing air over a mass of Japanese fireflies, *Luciola*, will start flashing and Ruedemann (1937) has reported a simultaneous flashing in American fireflies due to a pressure wave from exploding firecrackers.

The mechanism of flashing in the firefly is intimately connected with the structure of the organ (see p. 70). A mass of luminous cells on the ventral surface is backed by a "reflector" layer of cells containing minute crystals acting as a matt surface. The photogenic layer receives a rich tracheal supply, the air tubes passing through tracheal end cells and ending in the luminous cells themselves. Nerves also enter the photogenic layer. The consensus of opinion is that they are distributed to the tracheal end cells rather than the photogenic cells. On viewing the organ microscopically at night, one can see that the brightest light comes from the cytoplasm of the photogenic cell layer next to the tracheal end cells. Luminescence is undoubtedly intracellular in the firefly.

Lund (1911) showed that the flashing was not connected with the respiratory rhythm or any movement of dorso-ventral

PHYSIOLOGY OF LUMINESCENCE

muscles as had been previously supposed. An excised organ does not flash spontaneously because the ventral nerve cord, which carries the stimuli, has been cut, but an excised organ is highly irritable and responds to local stimulation by a local flash of light. Lund found also that sudden increased air pressure applied to a dark excised organ will cause a flash. It appears as if the increased pressure had forced a valve, previously closed, and brought air in contact with the photogenic cells. Dahlgren (1917), has described such a complex valve mechanism in the tracheal end cell.

This theory of flashing in the firefly has been most popular. It was adopted by Gerretsen (1922), Creighton (1926) and Snell (1932) and is to be contrasted with the equally probable view that nerves go to the photogenic cells, stimulating them directly as in other luminous forms which do not possess tracheal end cells. The problem would then be the same as in *Noctiluca* and the tracheal end cell would have no special significance. In favor of this view is the fact that tracheal end cells are widespread in insects, serving as a general intermediary structure between tracheae and tissue, not at all restricted to light organs alone. Indeed, Bongardt (1903) described tracheal end cells in the larvae of *Phosphaenus hemipterus*, that do not flash rapidly.

Against the valve theory is the fact that tracheal end cells are absent in *Pyrophorus* (Geipel, 1915) and most glow-worms, which do not flash, but whose light appears slowly, increases to a maximum, remains for some seconds and then slowly disappears. Fig. 79 is a photocell record of *Pyrophorus*, showing the slow development but rhythmic character of the light which probably represents volleys of nerve impulses to the organs. Dubois (1886) thought that muscles let blood containing luciferase to the organ in rhythmic pulses but this view is certainly not correct. Heinemann (1886) believed the rhythm was due to breathing movements but these can be observed to be much slower than the rather rapid fluctuations of light shown in the figure (Harvey, 1931).

A third theory of flashing in fireflies has been presented by Maloeuf (1938) who believes that nerves and fluid-filled tracheoles both enter photogenic cells. Nerve stimuli cause production of metabolites (lactic acid) whose osmotic pressure draws fluid from the tracheoles, thus allowing oxygen to enter the cell and cause luminescence. Oxidation of the lactic acid lowers the osmotic pressure and fluid again fills the tracheoles. Maloeuf finds that injecting hypertonic solutions into the abdomen results in a continuous glow, but isotonic or hypotonic solutions do not. The continuous glow at low oxygen tensions is attributed to accumulation of metabolites, at high oxygen tension to continual diffusion of oxygen.

Apart from histological examination of the organ, attempts to analyze the mechanism of flashing in the firefly have involved study of the effect of various drugs and conditions known to act in a definite manner on smooth muscle cells. Such an analysis is questionable since it is not certain whether contraction of a muscle valve mechanism closes the tracheal ends or opens them by working against an elastic ring on the tracheole (Creighton, 1926), or by contraction injects air into the photogenic cells (Dahlgren, 1917; Gerretsen, 1922).

Gerretsen, for example, finds three stages in the effect of chloroform vapor on *Luciola vittata*: (1) reversible abolition of the ability to flash (2) reversible constant luminescence of the organ (3) irreversible quenching of the constant light. The first stage represents tonic contraction of tracheal end cells,[2] shutting off oxygen, the second, secondary relaxation of tracheal end cells admitting oxygen continuously, the third, nar-

---

[2] Perhaps narcosis of ganglia sending out rhythmic stimuli.

cosis of the luminescent oxidations. Emerson (1935) found a similar effect of ether on *Photuris pennsylvanica*.

Creighton (1926) injected adrenalin into fireflies and found that they luminesced continually. He attributed this to the tonic contraction of the tracheal end cell mechanism, comparable to smooth-muscle contraction in vertebrates. Histological examination showed the tracheoles wide open in adrenalin treated, hardly visible in control fireflies.

Snell (1932) studied the effect of oxygen tension, by photoelectric cell recording of flashes. In pure oxygen (760 mm. Hg), as observed by a number of workers, the intact firefly organ glows steadily. If the partial pressure of oxygen is lowered, the same continuous luminescence occurs until about 300 mm. when flashing begins and is normal until the low tension of 25 mm. is reached. Then the duration of the flash is prolonged more and more and at 4 mm. a steady faint continuous glow appears, going out only at very low pressures. Between 152 mm. oxygen (air) and 25 mm. oxygen, the duration of a flash is the same, 0.14 second, while the intensity falls only from 75 to 50 units (Fig. 80).

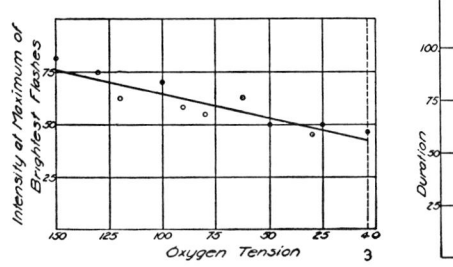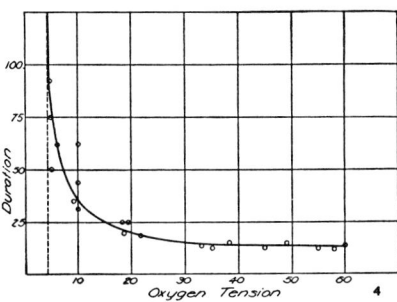

Fig. 80 Relation between the light intensity (left) and duration (right) of complete flashes of a firefly at different oxygen tensions, measured in mm Hg. Note that the flash does not become much brighter as the tension is increased from 40 to 150 mm Hg. The duration is also constant until low oxygen pressures, when the organ glows continuously. After Snell.

If the flash is controlled by nerve stimulation of tracheal end cells we should expect the opening of the tracheal valve to admit a certain amount of gas to the organ. The total amount of light in a flash would then depend on the amount of oxygen in the gas and should be much less, only one-sixth, at 25 mm. than at 150 mm. On the other hand, if direct stimulation of photogenic cells results in a flash, we should expect about equal total light over a medium range of oxygen tensions, in agreement with the general observation that luminescence of bacteria or of *Cypridina* luciferin-luciferase mixtures is independent of oxygen tension until the tension falls to very low values. Actually the total light in the flash of a firefly (intensity × duration of flash) is only slightly less (33 per cent less) over a wide range of tension. Although Snell did not interpret his results in this way they seem to point against a valve mechanism.

The meaning of continuous luminescence in pure oxygen is unknown. This tension may be high enough to allow diffusion past the valve or pure oxygen may paralyze some part of the nerve mechanism.

The continuous faint glow at low tensions is to be expected on the valve theory. We may suppose that the valve mechanism is asphyxiated or that reflexes admitting as much oxygen as possible to the whole animal would be called into play. It is interesting to note that at low tensions a sudden increase in oxygen tension results in a flash of light—a pseudoflash—showing that there is no barrier to entrance of oxygen and that oxygen supply *can* regulate flashing. Clogging the tracheal system with water also prevents luminescence. A similar flash was observed by Lund (1911) when the pressure around an *excised* organ in air was suddenly increased by squeezing a bulb. Although reflex flashing of *whole* animals from pressure waves is known, in this experiment the pressure must have directly forced more air into the tracheoles.

PHYSIOLOGY OF LUMINESCENCE

An interesting case of interference with the synchronous volley of impulses giving a flash is observed with fireflies bitten by spiders. The poison breaks up orderly discharge of nerve cells and the organ resembles a spinthariscope in that many isolated points of light momentarily appear scattered over the surface (Wood, 1939). Kastle and McDermott (1910) report a similar phenomenon after injecting strychnine, and have discussed the effects of other substances, as well as the experiments of the older workers.

It will be noted that the physiology of flashing is in a very unsatisfactory state, despite the abundance of fireflies and the fascination of the problem.

NARCOSIS

Narcosis or anesthesia, which may be defined as a temporary inhibition of some physiological activity, may involve a wide range of processes, such as cell-division, responses to stimuli of various kinds, light-production, mental activity, etc. Its chief characteristic is that the condition is reversible—that is, that normal processes are resumed on removal of the narcotic or anesthetic. Narcosis defined in this way may be caused by various means—lack of oxygen, the constant electric current, change of temperature, and many chemical substances. Among the latter are potassium cyanide, the salts of magnesium, calcium and other metals, as well as the more common lipoid soluble narcotics—chloroform, ether, alcohols, urethanes, barbiturates, etc. It is only lipoid soluble narcotics that will be considered here.

Narcotics may affect either the light emitting reaction or the stimulating mechanism. The former is best studied in bacteria, or in a luciferin-luciferase mixture itself. Numerous investigations (Ballner, 1907; Harvey, 1915; deCoulon, 1916; Taylor, 1934, 1936; Johnson, 1938; van Schouwenburg, 1938; Johnson and Chambers, 1939) have shown that lipoid soluble narcotics in the proper concentration will reversibly diminish luminescence of luminous bacteria without greatly affecting oxygen consumption. Further increase of the concentration of narcotic affects the general respiration also. Johnson, van Schouwenburg and von der Burg (1939) have demonstrated that ethyl urethane greatly reduces the characteristic flash of luminescence when air is admitted to bacteria kept in absence of oxygen. This is conclusive proof that urethanes suppress luminescence by a direct effect on the light emitting reaction, probably by covering the surface of luciferase.

McKenny (1902) noted an extraordinary development of tolerance to ether. The light of *Bacillus phosphorescens* disappeared in 0.5 per cent ether but the organisms grew luxuriantly and if continually treated and grown in 0.5 per cent ether, later developed luminescence and an ether resisting strain appeared.

The author has observed (1916) that a glowing tube of *Cypridina* luciferin and luciferase has its luminescence reversibly dimmed by addition of butyl alcohol and Taylor (1934) found the same for ethyl urethane. It is the luciferase which is affected, presumably by adsorption of the narcotic. In this case there can be no complication connected with cell surfaces or stimulation.

In *Noctiluca* the mechanism of stimulation is interfered with by anesthetics. E. B. Harvey (1917) found that in proper concentration of chloroform, ether, butyl alcohol, chloretone, thymol and ethyl alcohol, "the effect of the anesthetic has been in all cases not to prevent light-production altogether, but to prevent a normal response—i.e. a flashing on stimulation. In all the effective concentrations, the animals under the anesthetic produced a faint steady glow, so faint in some cases that it is not noticeable unless the animals are pres-

PHYSIOLOGY OF LUMINESCENCE

ent in large number. When returned to sea water, if not left too long in the solution, the steady glow ceases and the normal response returns; this is therefore a reversible phenomenon and a true case of anesthesia. . . . If returned to sea water *after* the period of steady glow, the animals gave no response, the prolonged anesthesia causing death."

Theories of narcosis fall into two groups: those which consider the narcotic to act upon the cell membrane and those which consider the action to be directly upon the cell contents. Since luminescence in *Noctiluca* only occurs in presence of oxygen, the loss of ability to flash in ether might be due to prevention by ether of penetration of oxygen into the cell. A simple experiment showed that such was not the case, for no bright luminescence appeared when narcotized cells were broken up with sand so as to destroy their membranes and admit oxygen. They always gave a much weaker light than the control unnarcotized *Noctilucae*. The evidence from this experiment therefore argues against the oxygen penetration theory of narcosis, since the narcosis of *Noctiluca* takes place independently of the cell-membrane. As with luminous bacteria, the effect of the narcotic seems to be upon the oxidative process inside the cell.

In higher organisms also, narcotics affect luminescence. Since light production is usually a reflex depending on stimulation of definite sensory endings the physiological problem is to determine which link in the reflex arc is affected first—receptors—nerves—ganglia—synapses or photogenic cells themselves. The situation becomes most complicated in the firefly, already discussed.

LIGHT

If luminescence is useful to the animal possessing this power it can obviously have no meaning during the day. Most luminous animals shun the light

> "Like a glowworm in the night, the which
> hath fire in darkness, none in light."[3]

We might therefore expect to find, especially in marine forms stimulated to luminesce by agitation of the waves, an adaptive mechanism by which luminescence would be prevented in daylight, thus saving luminous material. Such an effect of light might be (1) a direct effect on the luminescent reaction or (2) an effect on the nervous mechanism of stimulation. There might also be a day-night rhythm of luminescence, i.e. luminescence at night but none in the daytime, even if kept in the dark continuously. Such diurnal rhythms are widespread in other activities than luminescence (see Welch, 1938).

A number of luminous animals have been described in which a day-night rhythm occurs but some of these reports may be erroneous due to incomplete dark adaptation of the observer's eyes. Change from light to dark vision may take half an hour or more. Massart (1893) found a day and night periodicity of luminescence in Noctiluca, Zacharias (1905) in Ceratium, Kofoid (1921) in dinoflagellates, B. Moore (1908) in some animals of the plankton, presumably copepods, and Neidermeyer (1911) in Pteroides griseum. Parker (1920) observed some inhibition of luminescence of the pennatulid, Renilla, by sunlight and a partial day-night rhythm. At night its phosphorescence could be reduced by exposing it to light and by day could be developed by putting it in the dark. These observations have been confirmed by Sumner.[4] Renilla can luminesce in 2 to 3 minutes in a dark room after sunlight but the light becomes much brighter in an hour. Crozier (1920) found that the balanoglossid worm, Ptychodera, would maintain for eight days a clear-cut nightly rhythm of luminescence

---

[3] Shakespeare, Pericles, Act II, Scene 3, line 43.
[4] Private communication from Dr. F. B. Sumner to Professor G. H. Parker.

while remaining continually in the dark. Heymans and A. R. Moore (1923) observed similar behavior in *Pelagia noctiluca*, finding that these medusae show no general luminescence in the daytime even when they have been kept continuously in a dark room. In a later paper A. R. Moore (1926) was unable to confirm the observations of Heymans and Moore on *Pelagia* and the author has also been unable to observe any inhibiting effect of strong sunlight on either the local or general luminescence of this form, although the exposure was from sunrise until noon of a cloudless day.

The author's experience (1926) has not confirmed the widespread occurrence of day-night rhythms of luminescence. I have never yet observed an animal which would not luminesce during the daytime, provided it had been kept in the dark for a few hours previously. The copepods at Naples, *Noctiluca miliaris* at Misaki, Japan, and Friday Harbor, Wash., *Balanoglossus minutus* at Naples, and *Ptychodera bahamensis* at Bermuda luminesce during the day, in fact immediately after being brought to a dark room.

Other forms in which the *whole animal* had been exposed to sunlight and showed no inhibition of luminescence when tested immediately in a dark room were as follows: the ophiurian, *Amphiura squamata*; the annelids, *Chaetopterus variopedatus* and *Microscolex phosphorea*; the radiolarians, *Thalassicolla nucleata, Colozoum inerme* and *Sphaerozoum punctatum*; luminous bacteria; a luminous fungus (*Panus stipticus*); the pennatulids, *Pennatula phosphorea, Cavernularia haberi*, and *Ptylosarcus sp.*; the ascidian, *Pyrosoma sp.*; the ostracod, *Cypridina hilgendorfi*; the medusae, *Mitrocoma cellularia, Phialidium gregarium, Stomatoca atra*, and *Aequora forskala*; a squid, *Watasenia scintillans*; a fish, *Monocentris japonica*; and the fireflies, *Luciola parva* and *viticollis*. Panceri

(1872) noted no effect of sunlight on the luminescence of *Pyrosoma* or of *Pteroeides*.

Since Allman's (1862) original observation that ctenophore luminescence is inhibited by light, many confirming studies have amply substantiated his findings, Panceri (1872), Peters (1905), Harvey (1921, 1925), A. R. Moore (1924, 1926), Heymans and Moore (1925). Ctenophores give no luminescence on mechanical stimulation after exposure to sunlight until they have been in the dark for 5 to 40 minutes. Part of the inhibition is to be attributed to nerve inhibition, but squeezing the animal through cheesecloth after sunlight exposure also results in only a faint luminescence. This "extract" does not give the usual bright starry luminescence when poured into fresh water, showing that light actually destroys photogenic substance and does not merely affect the irritability of photogenic cells. The extract regains its power to luminesce to some extent in the dark but its ability to luminesce is not nearly as great as a similar extract prepared from dark adapted ctenophores. Small fragments of the luminous canals from light adapted animals become brightly luminous if kept in the dark but a centrifuged extract containing only photogenic granules does not. It appears that only intact cells can again recover the power to luminesce after light exposure, although this is difficult to prove with certainty.

In ctenophores, we have seen that oxygen is not necessary for luminescence[5] but an extract or a fragment of the animal loses its ability to luminesce in the light, whether oxygen is present or absent (Harvey, 1926). The "light adapted" fragments recover ability to luminesce in the dark only if oxygen is present, not in its absence. This suggests that the photogenic material of this form contains bound oxygen which is used up

[5] Dr. A. M. Chase informs me that oxygen is necessary for stimulation to luminescence of whole ctenophores.

PHYSIOLOGY OF LUMINESCENCE

in luminescence or after exposure to light. Consequently no reformation of photogenic material can occur in the dark if oxygen is absent (Harvey and Korr, 1938).

The violet and near ultraviolet end of the spectrum is most active in suppressing luminescence in ctenophores. If we expose to ultraviolet light without the visible, making use of nickel oxide ultraviolet filters, it can be observed that during the destruction of photogenic material no luminescence occurs. If the ctenophores have been previously stimulated, the end products of luminescence will fluoresce in the ultraviolet. Such animals are striking objects, for the eight canals are outlined with fluorescent light much like the true chemiluminescence of the ctenophore. The fluorescence disappears as oxidation products diffuse away (Harvey, 1925).

Peters (1905) noted that mechanical stimulation of the whole comb-jellies accelerates the appearance of luminescence in darkness after previous exposure to light and that the inhibition of luminescence is roughly proportional to the intensity of light which has previously illuminated them. It is found that for inhibition of a whole animal, a certain illumination acting for a certain time is necessary. This quantity is 4,776 meter candle minutes[6] for *Mnemiopsis* (Moore, 1924), 50,000 m.c.m. for *Beroë forskala* (Heymans and Moore, 1925) and 1,167 m.c.m. for *Cestus veneris* (Moore, 1926). Illumination × time is a constant quantity for inhibition, varying with the species. It is greater for photogenic material than for reflex stimulation to luminescence. Only that part of an animal directly exposed to light is inhibited. The rate of disappearance of photogenic material proceeds as a first order reaction and is independent of temperature, as are most photochemical processes which depend on absorption of light.

[6] See definition of units in Chap. VI.

LIVING LIGHT

In many animals the effect of light is chiefly on a nerve mechanism. Heinemann (1872) observed inhibition of the light of the *Pyrophorus* beetle during the day. Some *Pyrophorus noctilucus* sent to me from Cuba showed no inhibition of the teased organ in bright illumination but if one of the two luminescing thoracic organs in the intact animal was exposed to sunlight or the carbon arc light (passing through water) for 2 minutes, it could be observed afterwards that the exposed organ glowed less strongly than the other in the dark. We are evidently dealing in these cases with no true inhibition of luminescence of luminous material but with an effect of light on the local mechanism of stimulation in the organ itself.

Fireflies also behave like *Pyrophorus*. I have observed no inhibition of the luminous organs of *Photuris pennsylvanica* removed from the body and mashed on a slide (so that it luminesced continuously) after 4 minutes' exposure to 15,000 foot-candles.[7] However, when the luminous organ (alone) in the intact animal was directly exposed to the above light, the normal flashes of the animal as well as the flashes from electrical stimulation of the nerve cord in the thorax were inhibited and remained so for some seconds after the light was screened. Illumination of the eyes by 15,000 foot-candles, while keeping the luminous organ in the dark, does not inhibit the flashing but Gerretsen (1922) noticed that illuminating the eyes of *Luciola vitticollis* did stop flashing.

Buck has studied *Photinus pyralis* with great care. He finds that a change from bright to dim light will start flashing at any time of day. Change from darkness to dim light induces flashing provided the fireflies have been in darkness for 24, 48, 72, or 96 hours but not if in darkness for 12, 36, 60, or 84 hours. There is thus an inherent diurnal periodicity in periods of flashing, manifest over a four-day period in complete darkness, but

[7] See definition of units in Chap. VI.

PHYSIOLOGY OF LUMINESCENCE

these periods need not coincide with any actual time of day. The flashing itself is always induced by a definite light intensity.

Skowron (1926) noted that earthworms are less readily stimulated to emit their luminous secretion in the day than at night, presumably an effect on reflex control. Whole *Cypridinae* can still secrete luminous material in a bright light but light does have an effect on crude luminous extracts of the animal, accelerating the oxidation of luciferin, as described in Chapter IV, p. 139. The luminous secretions of *Pholas* or of *Heteroteuthis* are not affected as are *Cypridina* extracts.

Luminous bacteria are not affected by strong light unless it contains far ultraviolet or heats the organisms. Suchsland (1898) exposed bacteria protected by a water filter to 11 hours of sunlight with no effect on luminescence. Visible light does not cause even a temporary dimming, as I have determined (1925) by fixing the bacteria on a rotating wheel in such a way that they are illuminated one moment and examined in the dark the next moment, 1/200 second later.

In the far ultraviolet region, weak radiation may prevent growth but not luminescence, while a greater intensity affects luminescence also (Beijerinck, 1915). This results in delayed effects on luminescence—no change until some time after radiation when a gradual diminution in light intensity begins. The radiation does not affect the culture medium but brings about changes in the bacteria. Gerretsen (1915, 1920) occasionally found that after 30 minutes' exposure, the radiated half of an agar plate covered with bacterial colonies might be brighter than the unradiated but the luminescence soon faded. He thought that the ultraviolet light liberated oxygen from the medium. Another possible explanation is that the radiated organisms could not grow and consequently nutrient material was not used up nor were waste products produced as rapidly as in non-radiated regions.

The lethal effects of ultraviolet are like those observed with any type of bacteria. With sublethal doses colonies grow much more slowly. Gerretsen was unable to obtain non-luminous mutants by ultraviolet light comparable to those produced by Beijerinck (1912) after growth at high temperatures.

RADIUM RAYS

Omeliansky (1911), Beijerinck (1915) and Zirpolo (1920) have all studied effects of radium rays, the first finding a diminution of luminescence, the last an increased intensity. Perhaps these contrary results are to be explained in the same manner as the effect of ultraviolet light, since Beijerinck found first a prevention of growth by radium rays without affecting luminescence, later a decline in luminescence and the death of the cell. Vormolen's (1918) work has dealt with mesothorium and polonium while Zirpolo (1921) has studied uranium and thorium salts.

TEMPERATURE

Temperature effects on luminous animals are in part connected with the reaction emitting the light and in part with the mechanism of stimulation. In the primary effect, rise in temperature from a minimum for luminescence increases the light intensity to an optimum above which the intensity falls. A maximum temperature for luminescence is a somewhat variable quantity since it depends on the time of heating. A luminescence may be extinguished by heat and return if the organism is quickly cooled.

The primary effect is unobscured in bacteria and the relation between luminescence and temperature has been worked out by Morrison (1925), Root (1932) and Akabane (1938). Root found that the temperature coefficient (rate at 20°/rate at

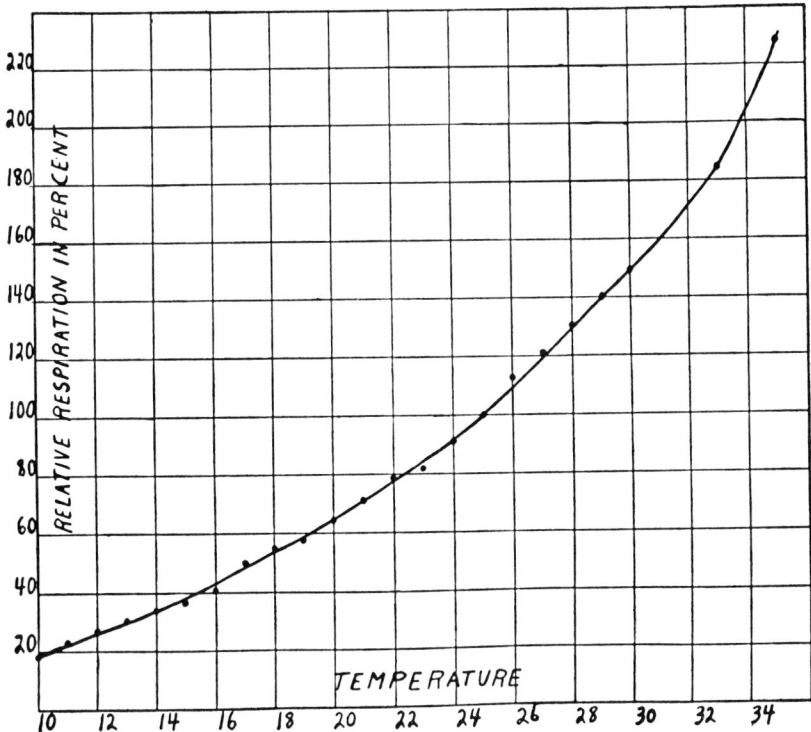

FIG. 81 Relation between oxygen consumption (vertical) of a freshwater luminous bacterium, *Vibrio phosphorescens* and temperature (horizontal). Note that the oxygen consumption is still increasing at the relatively high temperature of 34°C. After Root.

10° C.) for total respiration (Fig. 81) was 3.6 while that for luminescence intensity was 5 (Fig. 82). Light can just be seen coming from certain luminous bacteria at − 11.5° C. and + 34° C. (Harvey, 1913). Their resistance to liquid air temperatures has already been noted in Chapter II. Zirpolo (1933) found they could luminesce again after cooling to − 269.5° C. Beijerinck (1912) observed that luminous bacteria would grow although they did not luminesce at high tem-

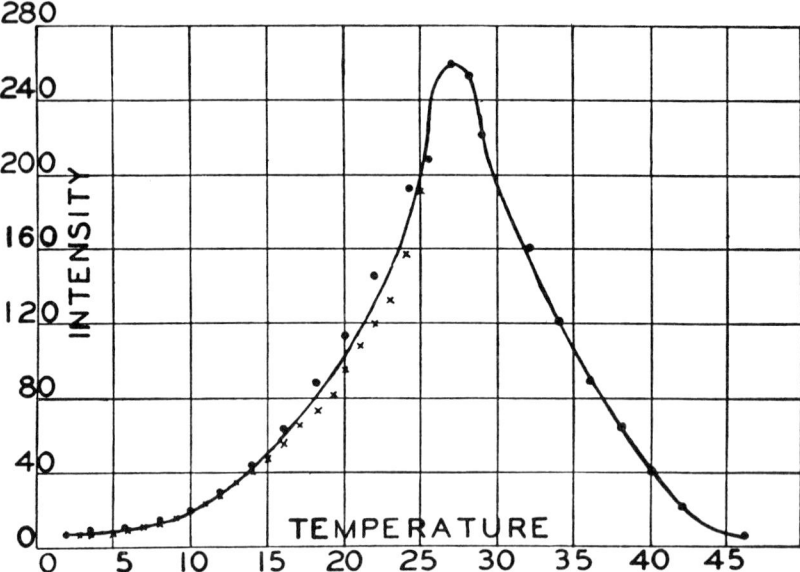

Fig. 82 Relation between luminescence intensity (vertical) of a freshwater luminous bacterium, Vibrio phosphorescens, and temperature (horizontal). Note the optimum temperature at 27°C. After Root.

peratures, and that frequently non-luminous mutants would arise that grew without luminescence at low temperatures.

In Noctiluca the effect of temperature on stimulation to luminescence (normal response) and on the light reaction itself (primary effect) can be readily distinguished since at both low and high temperatures the mechanism of stimulation is affected first. To quote from E. B. Harvey (1917):

"With increase of temperature up to 42° or 43°C., noctilucas give a normal response. From this point to 48° or 49°, a steady glow is given and then the light goes out completely and there is no recovery if cooled immediately. With decrease in temperature, the animals flash more than normally until the temperature reaches 5° to 0°, when they give a constant glow. If kept only a few minutes at 0°, they will recover on warming and again give a normal response; but if kept at 0° for 15 minutes they do not recover."

SALTS

Aquatic luminous organisms are almost exclusively marine but the luminescence itself is not immediately dependent on salts. Marine luminous bacteria can luminesce for a long time in pure isotonic sugar solution. Fresh-water luminous bacteria will live in distilled water although it is doubtful if they would grow indefinitely without salt.

Many experimenters have studied the growth of luminous bacteria with different salts in the medium. They are often remarkably independent of the kind of salt, provided the osmotic pressure is maintained, but the detailed results of these researches are so conflicting that it is not possible to draw any very definite conclusions. The investigations have been mostly carried out by growing the organisms in culture media whose sodium chloride has been replaced by other salts (McKenney, 1902; Molish, 1904, 1912; de Coulon, 1916; Gerretsen, 1920; Zirpolo, 1919-1923; Richter, 1926, 1928; Fuhrmann, 1932; Takase, 1939). Both Molish and Fuhrmann found growth and luminescence in potassium chloride, while McKenney, de Coulon and Gerretsen found the opposite. Both Molish and de Coulon found no ill effects of calcium chloride, while McKenney and Gerretsen observed no glow or luminescence in this salt. All agree that magnesium chloride is harmless. This will give an idea of the conflicting results with what may be termed physiological salts. There is general agreement that such unusual substances as sodium bromide or nitrate may sustain growth and luminescence but sodium and potassium iodide again have given contradictory results. Heavy metal salts are often toxic in minute concentration.

The discrepancies in these growth studies may be due in part to different organisms used and in part to lack of care in removing traces of other salts and in maintaining the proper reaction of the culture medium. Some of the salts are hydrolytically dissociated and these bacteria are particularly sensitive to changes in P_h. They do best on a well buffered nutrient medium.

The most important point is the independence of luminous bacteria of salt antagonism. The author (1915) has washed a suspension of luminous bacteria many times in pure sodium chloride containing no trace of calcium without any diminution of luminescence or other ill effects. In pure potassium chloride the light was faint after one hour and disappeared immediately in isotonic calcium and magnesium chlorides.

The most sensitive method of studying salt effects is to measure respiration and luminescence of the bacteria. Johnson and Harvey (1938) found the general respiration of well washed suspensions of *Achromobacter Fischeri* to be practically the same in isotonic sodium phosphate, sea water and pure sodium chloride, but diminished in isotonic potassium chloride, sucrose, magnesium chloride and was practically abolished in calcium chloride. Luminescence is more sensitive than respiration. It is almost quenched in calcium and magnesium chloride, greatly affected in potassium chloride and somewhat affected in sucrose.

The results of Clarens (1938) are not entirely in agreement. He also used well washed suspensions and studied respiration (but not luminescence) of *Micrococcus cyanophos* in different salt concentrations. Sodium chloride of 1.75 per cent containing a little glucose proved to be the optimum concentration. In glucose solutions of the same osmotic pressure but without salt, respiration was about cut in half. The sodium could be replaced by lithium, potassium or ammonium and

PHYSIOLOGY OF LUMINESCENCE

the chloride ion by bromide or iodide without great changes in respiratory rate, but the ions did have an effect that ran parallel to ionic size. Replacing one quarter of the sodium chloride with a phosphate buffer mixture ($P_h = 6.98$) increased the oxygen consumption, and adding a little magnesium chloride to this had a still further beneficial effect which was not apparent when magnesium was added to sodium chloride alone. The whole relation of bacteria and salts depends on the substrate and is decidedly complicated.

The behavior of marine luminous bacteria in diluted and concentrated sea water has been studied by Johnson and Harvey (1937, 1938). These bacteria do not swell in greatly diluted sea water but are apparently surrounded by a rigid membrane[8] which cracks (Hill, 1929), letting out some of the contents of the cell. At the same time the luminescence disappears. In moderately diluted sea water luminescence falls off practically in proportion to dilution but respiration may continue at the normal high rate in 50 per cent sea water. In concentrated sea water respiration falls off with increasing concentration but in twice concentrated sea water the luminescence may be little affected. The rather unusual relations are shown in Fig. 83.

Any animal requiring a balanced salt solution for maintenance of life will be affected in solutions of the pure salts of sea water and then its ability to luminesce impaired. *Noctiluca* is a good example. Any unfavorable condition is at once apparent since these flagellates then become insensitive to stimulation and glow steadily. This reaction is to be observed in diluted sea water (35 per cent). In cane sugar the response is normal for two hours and then the constant glow appears but addition of 10 per cent of sea water to the sugar

[8] This is probably the reason these bacteria are not cytolysed by saponin and the bile salts, which do not affect the luminescence.

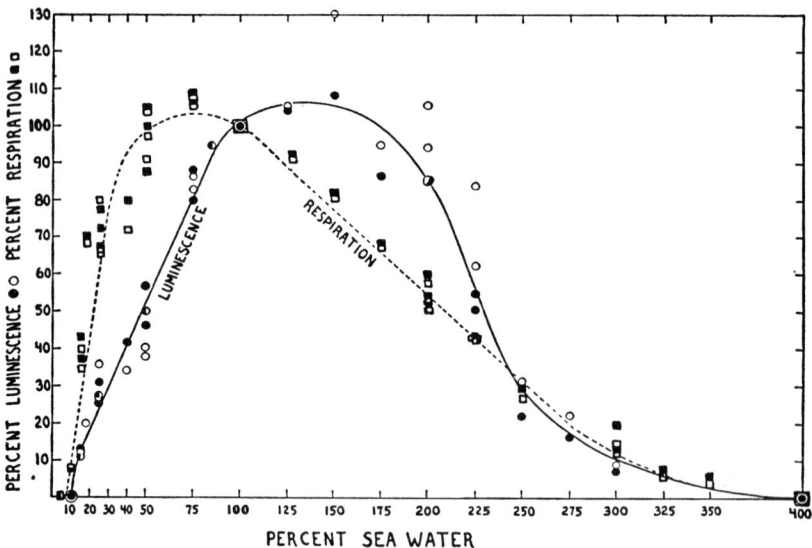

FIG. 83 The effect of diluted and of concentrated sea water on the luminescence (solid line) and oxygen consumption (dotted line) of the salt-water luminous bacterium, *Achromobacter fischeri*. After Johnson and Harvey.

solution is sufficient to maintain a normal response for days. The necessity for a small amount of salt is evident.

In pure isotonic sodium chloride the light response of *Noctiluca* on stimulation is rather weak and then a constant glow sets in but a trace of calcium chloride (the same proportion as in sea water) will preserve the normal light response for five days. As more of the salts of sea water are added in proper proportions the longer does the *Noctiluca* remain normal. In pure potassium and magnesium chlorides the normal response lasts about an hour while in calcium chloride the constant glow appears in two minutes. The contrast of *Noctiluca* and bacteria in regard to salt effect is most striking.

In more complicated forms, such as jellyfish or ctenophores, the life of the animal depends also on a balanced medium

PHYSIOLOGY OF LUMINESCENCE

and the necessity of various ions for maintaining irritability and conductivity along nerves which activate luminous cells has been studied by Moore (1924), Heymans and Moore (1924, 1925) and Hykes (1928). If a filter paper containing the luminous slime of *Pelagia* is prepared and placed in sea water, isotonic sugar, pure sodium chloride or pure magnesium chloride there is no light. In calcium, potassium, barium, strontium and lithium chlorides or magnesium or potassium sulphate, a luminescence appears. The meaning of these results, whether the effect is on cells or luminous granules, is not clear.

Luminescence of luminous substances alone, for example the light due to *Cypridina* luciferin and luciferase can occur in complete absence of salts but Anderson (1937) found that the total light emitted by a mixture of luciferin and luciferase might be three times greater in 0.5 m. sodium chloride as compared with water. It was noticeably greater in 0.001 m. sodium chloride than with no salt present. Potassium chloride, sodium bromide, potassium bromide, fluoride, nitrate and sulphate also increased the total luminescence in the order named while potassium sulphocyanide and iodide greatly decreased it. The results may be compared to the action of salts on the fluorescence of substances like quinine sulphate where the radiation is generally decreased. Such effects are totally different from the antagonistic ion behavior of *Noctiluca* or the action on the luminous slime of *Pelagia*.

HYDROGEN ION CONCENTRATION

All organisms require a definite P_h range in which to live, and luminous forms are no exception. Ammonia is frequently added to sea water to make its contained organisms glow, but too much quenches the light and it was early recognized that a few drops of acid would prevent the phosphorescence of the sea. Statements in the older literature are worthless since the hydrogen ion concentration was not known.

Vouk, Skoric and Klas (1931) give $P_h = 5.5$ to 9.0 as the limits of a marine luminous bacterium. Luminous bacteria respire and luminesce well between P_h values of 5.9 to 8.3, a normal range (Clarens, 1938), and Eymers and van Schouwenburg (1936) were able to photograph the luminescent spectra in phosphate buffers between P_h 5.3 to 8 while van Schouwenburg (1938) found little change in oxygen consumption between P_h 5.7 to 8. Hill (1928) has pointed out that the beneficial effects of moulds growing on culture media is connected with the neutralization of acid formed by the bacteria and has studied (1929, 1932) various culture media and the conditions for luminescence in the lower fatty acids and their sodium and ammonium salts.

RESPIRATORY POISONS

In recent years a number of specific substances have been found to inhibit certain steps of the respiratory mechanisms of all cells. The effect of indifferent narcotics, particularly active in suppressing luminescence of bacteria, has already been considered (p. 172). Cyanide, carbon monoxide, pyrophosphate, arsenite, iodoacetate and fluoride have also been studied, using luminous bacteria.

Cyanide is undoubtedly the most important oxidative poison because of its effect on the widely distributed cyanide sensitive iron respiratory catalysts, and its harmlessness for dehydrogenases. All luminous bacteria are cyanide sensitive for total respiration but much less so for luminescence intensity. The latter is only slightly affected and addition of cyanide is a very useful means of preserving a long lasting luminescence of bacterial suspensions, since it reduces the rapid respiration and consequent utilization of dissolved

PHYSIOLOGY OF LUMINESCENCE

oxygen. The author observed the relatively slight effect of dilute cyanide on the luminescence of bacteria in 1915 but the results were only published in 1917. Later (1923) the bacterial organisms in the fish, *Photoblepharon*, were found to be affected by cyanide in the same way as ordinary luminous bacteria. DeCoulon (1916) noted that luminous bacteria treated with cyanide took a long time to reduce methylene blue and that the oxygen consumption was very low. These results have been confirmed by all other workers (Taylor, 1932; Eymers and van Schouwenburg, 1937; Clarens, 1938). The fact that luminescence is reduced by cyanide, although only slightly, indicates very definitely that in bacteria the oxidations resulting in luminescence are partially linked with the general respiration.

This is in striking contrast to the effect of cyanide on luminescence extracts apart from cells. A glowing mixture of *Cypridina* luciferin and luciferase, extracts of the pennatulid, *Cavernularia*, and of the firefly (Harvey, 1917) as well as the luminous slime of the earthworm (Skowron, 1928) are quite unaffected by high concentration of cyanide. Even in *Noctiluca*, potassium cyanide of 0.004 to 0.005 m. concentration allows a normal response for some time, followed by a constant glow appearing sooner the more concentrated the cyanide (E. B. Harvey, 1917). It would be of great interest to know how cyanide affects the luminescence of the granules from the gland of the squid, *Heteroteuthis* and the fish, *Malacocephalus*, in order to decide definitely the status of these forms—whether the light is intrinsic or of symbiotic bacterial origin. Only in the latter case should cyanide decrease the luminescence.

Van Schouwenburg (1938) has made an extensive study of cyanide poisoning of luminous bacteria, particularly with reference to oxygen tension. As this is changed, the sensitivity to cyanide of both luminescence intensity and total respiration increases with increase in oxygen tension but the light emission is relatively much more affected at the higher tensions. Van Schouwenburg interprets this to mean that the ferment-haemin system is completely paralyzed by cyanide in pure oxygen and the respiration of the bacteria shifted to oxidations making use of the luciferin system as a hydrogen acceptor. The work of Johnson, van Schouwenburg and van der Burg (1939), who find that cyanide has no effect whatever on the flash of luminescence after bacteria are deprived of oxygen, gives the final proof that cyanide acts on their constant luminescence in a secondary manner and not on the luciferase itself.

As cyanide has no effects on anaerobic metabolism it is to be expected that it could exert no influence on the anaerobic redox potential of luminous bacteria and none was found by Korr (1935), who has made a careful and detailed study of various other respiratory poisons. The results indicate that if respiration and luminescence are affected, the anaerobic redox potential will also change, but not if the respiration and luminescence are unaffected.

Arsenite and *iodoacetate* both greatly reduce the general respiration and luminescence in small concentrations. Bromacetate is about half as effective as iodoacetate and chloracetate has no effect.

Fluoride and *pyrophosphate* on the other hand, do not inhibit respiration and luminescence even in high concentration. None of the above four inhibitors affect the light of *Cypridina*.

Carbon monoxide. Little work has been carried out to determine the effect of this gas on luminescence of luminous animals. Only luminous bacteria have been studied and Clarens (1938) observed no more effect on respiration and

PHYSIOLOGY OF LUMINESCENCE

luminescence in 95 per cent carbon monoxide than in the corresponding high concentration of nitrogen.

Deuterium oxide. Heavy water, added to 80 per cent, in a nutrient solution, reduces the respiratory rate to about half and greatly diminishes the luminescence (Harvey and Taylor, 1934); see Zirpolo, 1938.

RESPIRATORY ACCELERATORS

If a bacterium is supplied with ample food, there are few substances which will increase the respiratory rate and none that will increase the luminescence; with little food material, many substances can increase the low respiration and luminescence. Taylor (1932) noted especially the effect of food on bacteria suspended in sea water and found that addition of glycerine, glucose and peptone all markedly increased the rate of oxygen consumption of a *slowly* respiring suspension. He investigated a number of hormones, some of which increase metabolism of mammalian tissue cells, and found that certain ones increased the respiratory rate apart from serving as food material but most appeared to act merely as nutrient. Epinephrine was the only hormone to affect the luminescence, which was markedly decreased by low concentrations.

In order to study properly the effect of food, the washed cell technique must be used. Luminous bacteria are thoroughly washed with sea water until only a low "endogenous" respiration and luminescence remain. Then the material to be studied is added. The method has been extensively used by Johnson (1936, 1937, 1938), who has investigated a long list of sugars and related compounds. If the sugar can be utilized, the respiration and luminescence will immediately rise when added to the washed bacteria. Only three and six carbon sugars and their polyhydric alcohols were found to be oxidized by *Vibrio phosphorescens* and *Achromobacter fischeri*. An interesting phenomenon is the marked inhibition of oxidation of certain sugars by alpha methylglucoside, an effect not connected with the splitting off of methyl alcohol but presumably due to competition for the surface of the oxidative catalysts. Alpha methylglucoside is absorbed at the expense of the sugar. Consequently no oxidation of the latter can occur. Luminescence is also affected by this specific inhibitor.

Dyes whose rate of reduction by the cell and whose reoxidation by oxygen is not too slow, can greatly accelerate the respiration of cells. Their redox-potential must be positive to that of the cell. The author (1929) studied this value for luminous bacteria and found that under anaerobic conditions indigo disulphonate could be reduced but the monosulphonate could not. Korr (1935), in a carefully controlled investigation, measured the redox potential of anaerobic suspensions of Achromobacter fischeri directly with gold and platinum electrodes placed in the suspension. Under standard conditions, the potential regularly fell to $E_h = -.214$ volt, a value between that of the two indigosulphonates mentioned above. Dyes such as methylene blue are easily reduced and Taylor (1932) observed that the proper amount of methylene blue greatly increased the respiration of luminous bacteria in a non-nutrient medium and slightly increased the respiration in a peptone nutrient medium, in agreement with the general effects on cells of the hydrogen acception. The blue color of the dye prevents an accurate study of the corresponding effect on luminescence.

The nitrophenols and halogenphenols are particularly active respiration stimulants. Only one has been studied with luminous bacteria. Shoup and Kimler (1934) found that 2-4-dinitrophenol in neutral or alkaline solution greatly increased the respiration in small concentrations, without affecting luminescence. In larger concentrations the respiration and finally the luminescence was diminished. Regarded merely as cells, there is no reason why luminous bacteria should behave in an unusual way.

Finally the question may be raised as to what happens when a strain of bacteria suddenly becomes non-luminous. The problem is akin to that involving the origin of any mutant. Pigment forming bacteria sometimes become colorless; virulent strains become harmless. We may suppose that in dark or dim strains there is lack of some necessary constituent for formation of luciferin or luciferase. Doudoroff (1938) has tested this viewpoint by studying the ability of lactoflavin (vitamine B_2), the prosthetic group of the yellow respiratory enzyme, to restore luminescence in dim strains. He found that some of the semidark dissociates of bright luminous bacteria did have their luminescence increased by small amounts of lactoflavin whereas others did not. The results were not as regular as one might wish. So little is known of the precursor of luciferin, proluciferin, in any animal, or of the origin of catalysts in cells in general that it is rather premature to discuss the problem adequately. We can only suggest it as a very promising field for future investigation.

CHAPTER VI

PHYSICAL NATURE OF ANIMAL LIGHT

FUNDAMENTALS

LIKE sound, the measurement of light is complicated by its dual interpretation—in terms of an objective energy change which can be measured in definite physical units and in terms of our subjective appraisal of it. The word light is used not only for the visual sensation but also for the physical basis. Ultraviolet *light*, completely invisible, has no subjective meaning except a reminder that its nature is the same as that of light which is visible. Visible light occupies only a small part of the electromagnetic spectrum. Living light is still further restricted, since it includes only a part of the visible region.

We can completely designate any light if we describe the relative energy of different wave lengths which make up the beam.[1] Such a measurement is a spectral energy or spectroradiometric curve and includes both visible and invisible radiation. It describes the quality and intensity of the light emitted, but additional knowledge is necessary to tell how this light will affect the eye or any other receiving surface which it strikes.

Total radiation can only be measured by its heating effect but frequently visible wave lengths are present in such low intensity that it is impossible to measure them with any type of thermometer. Other instruments must be used. These instruments—the eye, the photographic plate, the photoelectric cell or the photronic cell—do not "see" the same wave

[1] A statement of the amount and kind of polarization must also be made.

PHYSICAL NATURE OF LIGHT

lengths equally well. Therefore, we must know the relative ability of these detectors to respond to different wave lengths. This is the spectral sensitivity curve. For the eye, such a relation is called the "visibility of radiation" curve. It expresses the relative ability of the eye to see wave lengths of equal energy.

Visibility differs for different persons and depends on whether their eyes are light adapted or dark adapted. A color-blind person would have a totally different visibility of radiation curve but the average of all normal people is shown in Fig. 84. Only if a photocell has the same response could it replace the eye for measurements involving human visibility. By the use of special filters it is possible to make the spectral sensitivity of a photocell very closely resemble that of the eye.

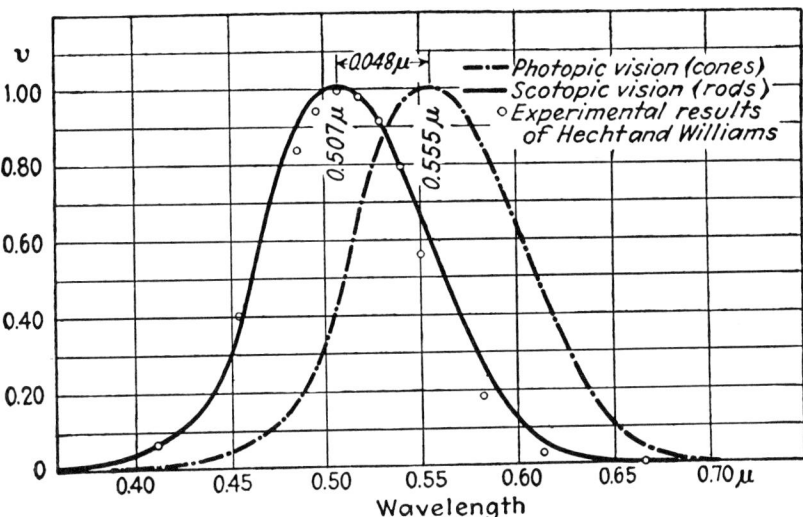

FIG. 84 Visibility of radiation curves for an average light adapted (broken line) and dark adapted eye (solid line). The relative visibility (vertical) is plotted against wave lengths (horizontal) containing equal energy. Note shift of the maximum toward the blue end of the spectrum during dark adaptation. Reproduced from Moon's *Scientific Basis of Illuminating Engineering*, by courtesy of McGraw-Hill Book Co.

From Fig. 84 it can be seen that the orange light of $\lambda = 0.61\mu$ or green light of $\lambda = 0.51\mu$ would appear half as intense to the light adapted eye as the yellow green, or twice as much energy in the orange or green would be necessary to produce as great an effect on the eye as in the yellow green. As the eye becomes dark adapted, the visibility of radiation curve shifts toward the blue. From a maximum in the yellow green at $\lambda = 0.555\mu$ there is a change to a maximum at $\lambda = 0.507\mu$ in the green. The dark adapted eye can see green better than yellow green and for this reason weak lights will appear more green than stronger ones of the same energy distribution. Weak lights of the same spectral composition therefore appear different in color if they differ much in intensity. This is known as the Purkinje effect.

The shift in sensitivity of the eye occurs in illuminations between 0.01 and 0.001 foot-candle[2] and represents a change from central cone vision (high intensities) to peripheral rod vision (low intensities). The central part of the retina, the fovea, lacks rods and this part of the eye becomes practically color blind at low intensities of light. Very weak lights appear grey with no color at all. It is clearly necessary then to take into account the behavior of the eye when any measurement of light is made.

A candle is, unfortunately, still the unit for measuring light intensity. No government bureau actually keeps a candle as standard for measurements but it does keep an incandescent light which has been calibrated in terms of the international candle, agreed to by many nations and based on a tallow candle. The candle is a physiological unit, since it depends on the ability of the human eye to detect light of visible frequen-

[2] A foot-candle = one lumen per square foot = 1.0764 milliphot (see p. 197).

PHYSICAL NATURE OF LIGHT

cies—luminous flux. It is no measure of the invisible frequencies and the total radiation or radiant flux.

A fundamental law of light intensities is the inverse square law, that the intensity varies inversely as the square of the distance, a purely geometrical consideration, dependent on a point source of light which spreads into space. If the point has unit intensity of one candle, it will emit in all directions 4π units of luminous flux or 4π lumens. The lumen is a unit for measuring amount of light. At a distance of 1 cm. from the candle each square centimeter will receive one lumen. On a surface at right angles to the light this illumination would be called a phot or a centimeter-candle. At one meter distance the illumination would be 0.0001 of a phot, sometimes called a lux or a meter-candle. The meaning of a foot-candle is obvious. If the surface illuminated by one phot reflected all the light of all wave lengths it would have a brightness of one lambert. The important distinction is between illumination and brightness, one denoting light received, the other light emitted or reflected. A grey and a white paper may receive the same illumination but they differ greatly in brightness.

All these facts, together with an additional one that describes what happens if the light makes an angle, θ, with the surface, can be summed up in the simple relations:

$$\text{Illumination in phots} = \frac{\text{Intensity in candles} \times \text{cosine } \theta}{(\text{Distance in cm.})^2}$$

and

$$\text{Brightness in lamberts} = \frac{\text{lumens}}{\text{square cm.}} = \frac{\text{candles} \times \pi}{\text{square cm.}}$$

Although mostly used for white light, these units apply also to colored lights which can be compared for brightness irrespective of color, although a visual comparison is not accurate if the colors are very different.

Although the lumen is a subjective unit, it can be converted into absolute units of energy by measuring the energy in one lumen of yellow green light of $\lambda = 0.555\mu$, the wave length of maximum visibility. This turns out to be 0.00038 calorie or 0.0016 watt-seconds. Since the watt is a unit of radiant flux, we can speak of the equivalent luminous flux as a lightwatt = 621 lumens. For comparison with electric light, 621 lumens per watt is generally accepted as the "mechanical equivalent of light"[3] at $\lambda = 0.555\mu$. It represents the maximum possible luminous flux from a watt. At other wave lengths, the lightwatts or the lumens per watt would be less, in proportion to the visibility of radiation curve.

Interest in the light of organisms from a physical standpoint has centered around questions of intensity, quality and efficiency. Living light is in no way different from light of ordinary sources, except in intensity and spectral content. It is all visible light, containing no infrared or ultraviolet radiation or rays which are capable of penetrating opaque objects. As first shown by Dubois (1886) for *Pyrophorus*, and confirmed by the author for bacteria and *Cypridina*, the light of animals is not polarized in any way but may be polarized by passing through a Nicol prism. Like ordinary light, animal light will also cause fluorescence and phosphorescence of substances, affect a photographic plate, cause marked heliotropism of plant seedlings (Nadson, 1903) and stimulate the formation of chlorophyll (Issatschenko, 1903, 1907; Molish, 1912).

[3] This expression is not strictly correct since the lumen is not an energy unit. All light units and photometric standards are defined in the *Transactions of the Illuminating Engineering Society* for 1932.

PHYSICAL NATURE OF LIGHT
INTENSITY OF ANIMAL LIGHT

A number of workers have estimated the light intensity of fireflies, since these forms appeal to human interest and are among the brightest and commonest of luminous animals. The values range from 1/1600 to 1/150 candle for *Pyrophorus* (Dubois, 1886; Langley, 1890; Pickering, 1916; Harvey and Stevens, 1928) while the flash of the firefly, *Photinus pyralis*, varies from 1/400 to 1/50 candle, the predominating values being around 1/400 candle. A continuous steady glow is sometimes obtained from this insect of the order of 1/50,000 candle (Coblentz, 1912).

The brightness of a luminescence is more important than the intensity in candles. Measurement of the light of the glowworm (larva of *Photuris pennsylvanica*) gave 14.4 millilamberts (Ives and Jordan, 1913) while the prothoracic organ of *Pyrophorus* had a brightness of 45 millilamberts (Harvey and Stevens, 1928). Moistened fragments of *Cypridina* glands, measured with an optical pyrometer, gave 2 to 16 millilamberts (Nichols, 1922; Dufford, Nightingale and Calvert, 1925). The ctenophore, *Mnemiopsis*, had a brightness of 0.3 millilamberts (Nichols, 1924). These values are about like those of bright fluorescences, which range from 1 to 35 millilamberts. To make such figures intelligible we can think of the blue sky as having a brightness of 1 lambert, paper properly illuminated for reading 4 millilamberts, and the luminous paint on watch dials 0.01 to 0.02 millilamberts.

The intensity of bacterial light is especially interesting. Lode (1904, 1906) found for bacterial colonies about 0.7×10^{-10} international candle per square millimeter of colony, while Friedberger and Doepner's (1907) value was ten times this. Lode calculated that the dome of St. Peter's at Rome, if covered with luminous bacteria, would give little more light than one common candle. The large luminous organ of the fish, *Photoblepharon palpebratus*, containing luminous bacteria, is equivalent in brightness to a surface illuminated with 0.0024 meter candles (Steche, 1909). I have observed (1925) a brightness of 23 to 144 microlamberts for well aerated bacterial suspensions in a vessel 2.7 cm. thick. After correcting for the absorption of light by other bacteria, a single luminous bacterium had an intensity of 1.9×10^{-14} candles.

QUALITY OF ANIMAL LIGHT

It is a fact that different luminous animals produce light of quite different colors as judged by our eye. A range of spectral tints has been described which extends from red to violet but "yellowish," "greenish," and "bluish" tints are commonest. Indeed, one or two animals possess several luminous organs emitting lights of different colors. This is true in the beetle, *Pyrophorus*, where the prothoracic organs are more greenish than the more yellow ventral organ, and especially in a South American firefly, *Phengodes* (= *Astraptor*), whose lights are red and greenish yellow. The deep-sea squid, *Lycoteuthis diadema*, produces lights of three colors, two shades of blue and red (Chun, 1910). The red light in the case of the squid may be due to a red color screen formed by the chromatophores, but in *Phengodes* no screen is present.

The first observations on the spectra of luminous animals were made by Pasteur (1864), who studied *Pyrophorus* and found a continuous spectrum unbroken by light or dark bands. Lankester (1868) discovered a similar continuous spectrum in *Chaetopterus insignis* and placed its limits from line 5 to 10 on Sorby's Scale (about $\lambda = 0.55\mu$ to $\lambda = 0.44\mu$). Young (1870) first recorded the limits of the firefly (*Photinus*) spectrum as a little above C ($\lambda = 0.656\mu$) to F ($\lambda = 0.486\mu$). Since then, a number of luminous forms have been examined and all are found to give short continuous spectra lying in different color regions.

Langley and Very (1890) placed the spectral limits of *Pyrophorus noctilucus* at $\lambda = 0.640$ to 0.468μ, a broad band chiefly in the green and yellow. But, "would the light not extend farther were it bright enough to be seen? . . . if the light of the insect were as bright as that of the sun would it not extend equally far on either side of the spectrum?" Langley and Very answered this question by making a solar spectrum from sunlight of the same intensity as that from *Pyrophorus* and comparing them in the same field of the spectroscope. The latter was very much shorter than the solar spectrum, showing that its length was not due to weakness of the red and blue rays but to their absence.

Ives and Coblentz (1910) were the first to photograph the spectrum of a firefly (*Photinus pyralis*) on plates sensitive to all wave lengths under conditions which would have recorded all radiations given off. They found the spectrum to extend only from $\lambda = 0.51\mu$ to $\lambda = 0.67\mu$. Another species of firefly, *Photuris pennsylvanica*, gave a spectrum extending from $\lambda = 0.51\mu$ to $\lambda = 0.59\mu$ (Coblentz, 1912). The *Photinus* light extends much farther into the red and it is easy to distinguish between *Photinus* and *Photuris* in the open field, merely by the reddish tint of the light of the former. These photographic records show conclusively that the color of the light of luminous animals is not a subjective phenomenon due to the low intensity of the light. It is real, an actual difference in spectral composition of the light emitted, not the result of colored pigment in the cuticle.

Several writers (Dubois, 1914; Fischer, 1888; Molisch, 1904) have reported that the light of luminous bacteria changes in color if grown on different culture media. Light which is "silver white" on dead fish becomes "greenish" on salt-peptone-gelatin media and more yellow on salt-poor media. Takase (1938) also found a difference in spectral energy distribution of the light of symbiotic bacteria from the deep-sea fish, *Coelorhynchus*, in salt and salt-free (sugar) solutions, but Eymers and van Schouwenburg (1936) could detect no difference in the spectral energy curves of *Photobacterium phosphoreum* connected with the age of the culture or external conditions such as temperature, salt content or hydrogen ion concentration.

A number of workers (Peron, 1804; Panceri, 1872) have noted color changes in the light of *Pyrosoma*. Polimanti (1911) describes the normal light of *Pyrosoma* as greenish, and states that as the animals die, or if they are kept at temperatures above the optimum, the light becomes more red. McDermott (1911), noticed that the light of fireflies placed in liquid air became decidedly reddish just before going out and on rewarming, the first light to appear was reddish followed by the normal shade at higher temperatures. I have frequently observed a more reddish color from luminous tissues of the firefly upon the addition of coagulants such as alcohol or on heating, and have noted that the light of *Cypridina* becomes weaker and more yellow at both low (0°) and high (50°) temperatures. This important point demands further investigation.

When Ives and Coblentz (1910) and Coblentz (1912) first photographed the spectrum of firefly light, they calibrated the plate so that its blackening could be interpreted in terms of energy. Coblentz and Hughes (1926) continued the work and plotted spectral energy curves for *Cypridina* and for luminous wood, whose light is due to the fungus, *Armillaria mellea*. The *Cypridina* curve is reproduced in Fig. 85. It will be noted that the energy distribution may be unsymmetrical and that it

PHYSICAL NATURE OF LIGHT

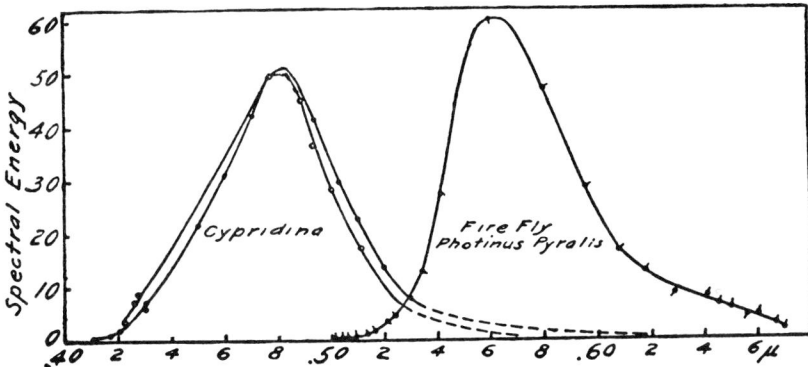

Fig. 85 Spectral energy distribution of the firefly and Cypridina light. Energy (vertical) is plotted against wave length (horizontal). Note the maxima in two very different spectral regions. After Coblentz and Hughes.

appears in very different regions, with maxima at $\lambda = 0.55\mu$ to $\lambda = 0.58\mu$ for fireflies and $\lambda = 0.48\mu$ for Cypridina.

The most recent work is by Eymers and van Schouwenburg (1936, 1937), who studied the spectral energy distribution in several species of bacterial light, in Cypridina, and also in a number of chemiluminescences, plotting the energy both as a function of wave length and of frequency.[4] The latter method of plotting brings out a few fundamental frequencies whose symmetrical broadening and combination gives rise to the complete curve. Certain frequencies may be common to a number of luminescences, as for example, 18,200 cm.$^{-1}$ which is found in luminous bacteria, in Cypridina, in the chemiluminescence and fluorescence of dimethyldiacridylium nitrate and in the fluorescence of lactoflavin and a fluorescent pigment produced by Pseudomonas putida. The lactoflavin fluorescence is made up of one symmetrically broadened frequency, 18,200 cm.$^{-1}$ This strongly supports the validity of an analysis

[4] Frequency is defined as the reciprocal of wave length expressed in cm. $\nu = 1/\lambda$. To obtain actual frequencies the figures must be multiplied by the velocity of light, 3×10^{10}cm. per second.

of unsymmetrical spectra as combinations of symmetrical ones. Wherever fundamental frequencies are found, a common configuration of the emitting molecule is to be expected. Takase's (1938) analysis of the light of symbiotic bacteria from the deep-sea fish, *Coelorhynchus*, showed fundamental frequencies at 21,200 cm.$^{-1}$ and 19,200 cm.$^{-1}$ On a cane sugar medium the 21,200 cm.$^{-1}$ is the more intense while the reverse is true on a salt medium. Eymers and van Schouwenburg's bacterial spectrograms for *Photobacterium phosphoreum, Ph. splendidum* and a fresh-water *Vibrio* all showed fundamental frequencies at 18,300 cm.$^{-1}$ and 20,400 cm.$^{-1}$ Further data will be necessary to establish the validity of these results. The bacterial curve is shown in Fig. 86.

EFFICIENCY

The area of these spectral energy curves gives the total radiant energy, the radiant flux. The visibility of radiation curve tells how each wave length affects the human eye. Correcting for this gives another curve, whose area is the luminous flux. The ratio of the two areas gives the luminous efficiency.[5] It is very high for the firefly, very low for an electric lamp, which radiates a great deal of heat. This is the basis for the widely heralded statement that the firefly is a 100 per cent efficient source of light, since its radiation is all in the visible. The statement is practically true if we refer only to the proportion of total radiation which affects our eye. It is far from true if

[5] The oldest method of estimating luminous efficiency was to divide the energy of visible wave lengths ($\lambda = 0.76\mu$ to $\lambda = 0.40\mu$) by the total energy radiated. A more correct method divides the energy of each visible wave length multiplied by visibility at that wave length, by the total energy radiated. This takes into account the ability of the eye to see the radiation and was called the reduced luminous efficiency. The ratio has also been called luminous efficacy, since it is not a comparison of energies but involves subjective measurement of the effect of light on the retina.

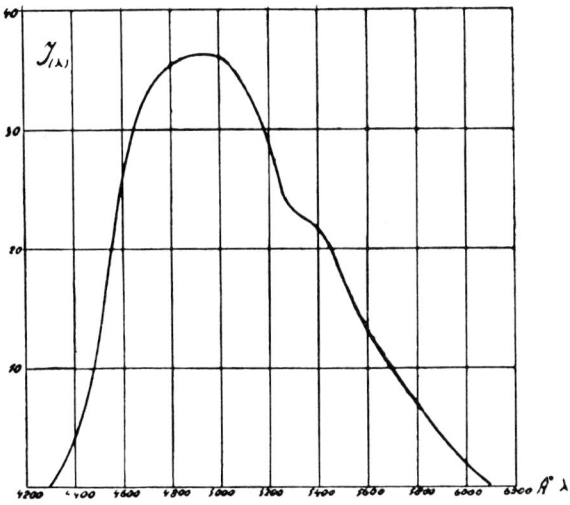

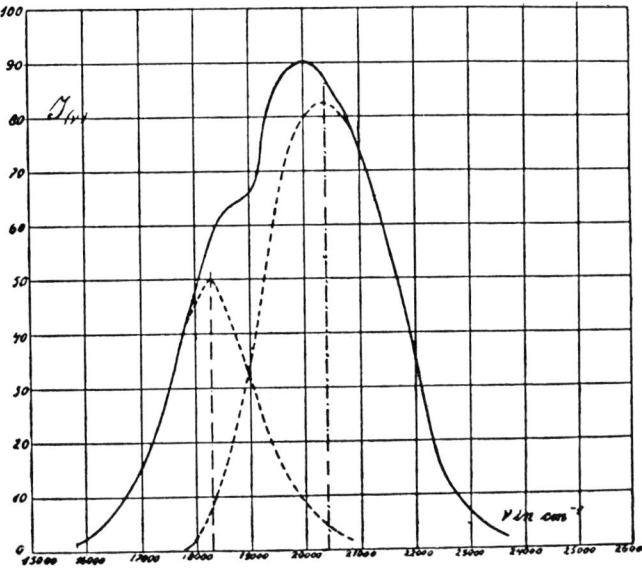

Fig. 86 Spectral energy (vertical) curves for *Photobacterium phosphoreum*, plotted as a function of both wave (above) and frequency (below), along horizontal. Note fundamental frequencies at 18,200cm^{-1} and at 20,300cm^{-1}. After Eymers and van Schouwenburg.

we take into account the efficiency of producing the radiation itself, that is if total efficiency is considered.

Early work was largely concerned with establishing the fact that all radiation is in the visible. The general method of study was to convert the radiation from a candle or from the firefly into heat on the blackened surface of a thermopile, bolometer or radiometer. The amounts of energy represented could then be compared in a common unit, the calorie. By proper screening, all rays except the visible light rays could be cut off from the measuring instrument and the amounts of energy represented in light and in total radiation be thus determined.

Dubois (1886) first studied this problem in *Pyrophorus* by the use of a thermopile and galvanometer and found a small amount of radiation from the luminous region in excess of that from a non-luminous region. It increased slightly during the flash of the insect but was very small.

A more careful study was made by Langley and Very (1890) with the bolometer. The radiation from *Pyrophorus* which affected their bolometer was shown to be due merely to the "body heat" of the insect, and it is largely cut off by a plate of glass which is opaque to all wave lengths of 3μ or longer, waves given off by bodies at temperatures below $50°C$. Langley and Very then compared the radiation from a non-luminous bunsen flame and the *Pyrophorus* light, interposing a plate of glass in each case to cut off the waves longer than 3μ, and found several hundred times more radiation in the case of the bunsen burner but, nevertheless, perceptible radiation from *Pyrophorus*. Later, Langley (1902) reinvestigated the radiation of *Pyrophorus* and could detect no heating whatever with the bolometer. "A portion of the flame of a standard sperm candle, equal in area to the bright part of the insects, gave under the same circumstances, a bolometric effect of such

PHYSICAL NATURE OF LIGHT

magnitude that had the heat of the insect been 1/80,000 as great as that from the candle, it would certainly have been recognized." As Langley and Very express it in the title to their paper, firefly light is "the cheapest form of light," not cheapest in the sense that we can reproduce it commercially at less cost than other lights, but cheapest in the sense that it is the most economical in the energy radiated.

Later Ives (1910), using a method based on extinction of phosphorescence of zinc sulphide, and Coblentz (1912) also, using a vacuum thermopile of platinum and bismuth, was unable to detect any infrared radiation from *Photinus pyralis*, but found that the temperature of this firefly is slightly lower than that of the air. He also measured the chitinous integument of the firefly to infrared and found it transparent except for a short band between $\lambda = 2.8\mu$ and $\lambda = 3.8\mu$.

Although photographs of the spectrum of firefly (*Photinus*) light, made with a quartz spectrograph, show that it extends only to the beginning of the blue, Forsyth (1910) reports ultraviolet radiation in luminous bacteria. He exposed a plate for 48 hours to the spectrum of bacterial light, dispersed by a quartz prism, and got a continuous band from $\lambda = 0.50\mu$ (the lower limit of sensitivity of the plate) to $\lambda = 0.35\mu$. All other studies have failed to confirm this observation. No ultraviolet radiation has ever been detected in the spectrum of any luminous animal.

Even though all radiation of the firefly lies in the visible range a correction of the spectral energy curve for visibility of radiation leads to lower luminous efficiency. A graphic method of presenting these relations in the firefly and incandescent lamp is illustrated in Fig. 87, where the visual sensitivity curve of the eye is used to correct the spectral radiation curve. The black areas represent the luminosity of the radiation which is some 0.5 per cent of the total in the lamp and around

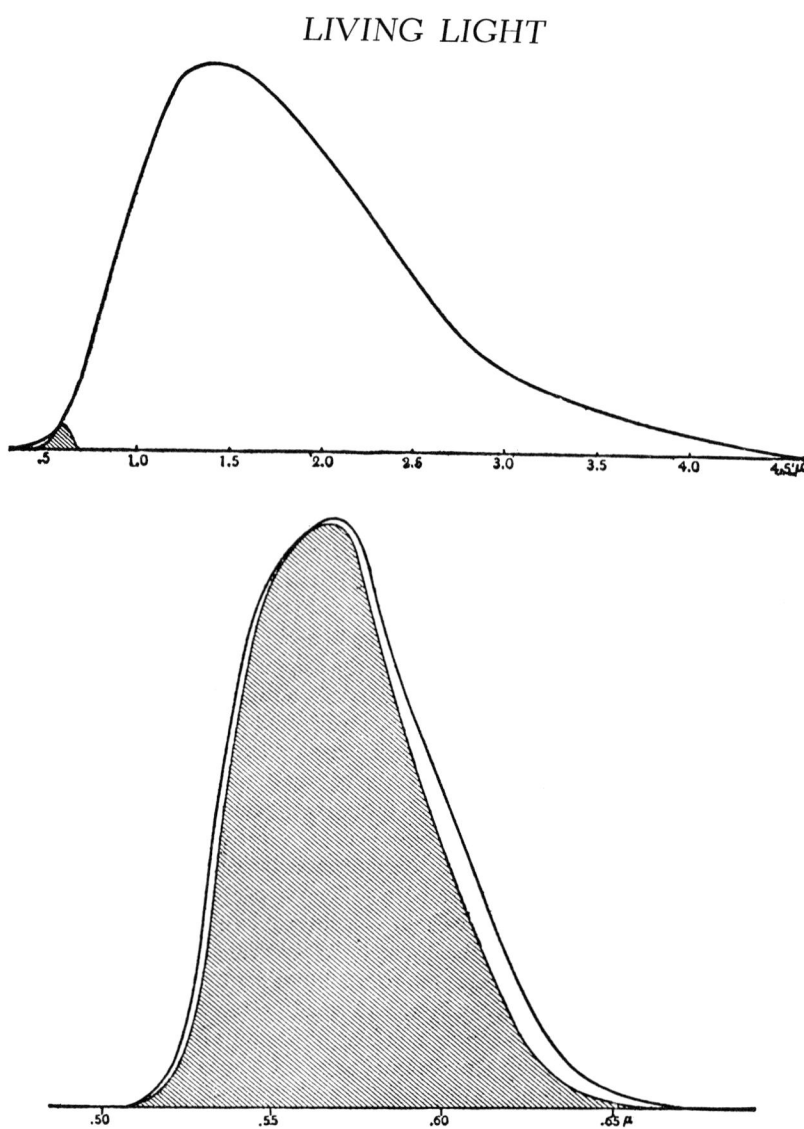

Fig. 87 Luminous efficiency of the carbon incandescent lamp (above) and the firefly (below). The shaded area represents the luminous flux and the total area the radiant flux in each case. The ratio of luminous to radiant flux is the luminous efficiency, 0.5 per cent for glow lamp and 95 per cent for the firefly. After Ives and Coblentz.

90 per cent in the firefly, depending on the species. Coblentz's (1912) values, using a different visual sensitivity curve for a partially light-adapted eye, gave the reduced luminous efficiency as 87 per cent for *Photinus pyralis*, 80 per cent for *Photinus consanguineus* and 92 per cent for *Photuris pennsylvanica*.

Although the firefly light has high luminous efficiency, it would be a trying one for artificial illumination, as all objects would appear of a nearly uniform green hue. Indeed the distortion would be even greater than with the mercury arc, whose objectionable green hue is so well known. Efficiency is not the only requisite for a lamp—convenience, cost and particularly color are of prime importance. The firefly has carried the striving for efficiency too far to be acceptable to human use; it has produced the most efficient light known but has produced it at the inevitable expense of range of color (Ives, 1922).

If we correct the spectral energy curve of *Cypridina*, recorded by Coblentz and Hughes (1926), for visibility of radiation, we obtain considerably less efficiency than in the firefly, 20 per cent. This is due to the blue color of *Cypridina* light, whose maximum is so far from the region of maximum visual sensitivity. The bacterial spectroradiometric curves of Eymers and van Schouwenburg (1937) show 45 per cent luminous efficiency, as illustrated in Fig. 88.

All the above high efficiencies tell nothing about the efficiency by which the chemical energy of oxidation of luciferin is converted into radiant energy or the over-all efficiency of the firefly regarded as a power plant for producing light. For any comparison with artificial illuminants such a measurement is absolutely necessary.

The energy for practical illumination originally came from combustion; it now comes from electrical energy. The blazing

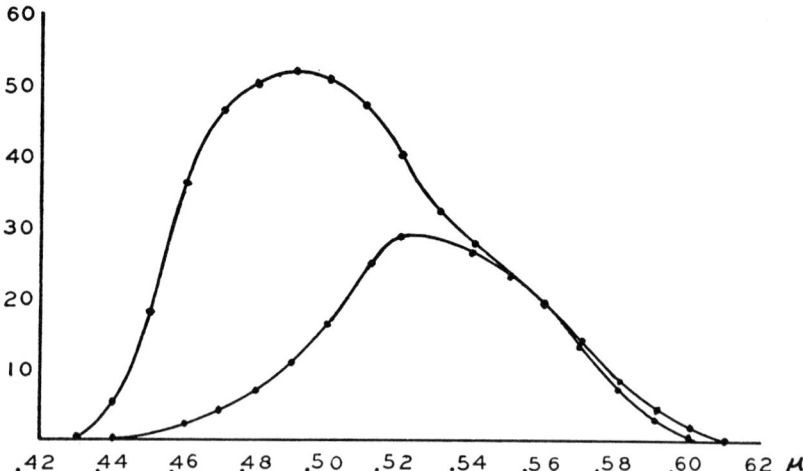

Fig. 88 Luminous efficiency of the luminous bacterium, Photobacterium phosphoreum, the ratio of the area under the lower curve (luminous flux) to the area under the upper curve (radiant flux), 45 per cent. The luminous flux has been drawn by the author in a curve of Eymers and van Schouwenburg for radiant flux.

torch, the candle, the oil lamp or gas light involve heating carbon particles or some other material (like a Welsbach mantle) by the heat of oxidation of the hydrocarbons in the flame. Without oxygen there would be no light. The total efficiency of such sources, for example a gas lamp, is made up of the product of two efficiencies. The first is luminous efficiency, the energy in the visible region evaluated according to the ability of the eye to see it, divided by the total radiant energy. This ratio we have just been considering in luminous animals. The second is radiational efficiency, the energy of the total radiation divided by the energy of combustion of the gas. If we start with the energy in the coal from which gas is produced, the efficiency of the conversion of the coal to gas must also be considered. We can then speak of the over-all efficiency of a gas lamp, from coal to visible light. It is this value about which little is known in luminous animals.

Similarly, in the case of electric light, when an incandescent lamp lights, coal is being burned in some power house. Some energy is lost in generation of current, as only about 33 per cent of the energy of the coal appears as electric energy at the lamp terminals. The energy of the coal is measured by the number of calories of heat produced when the coal is burned. As a pound of coal in burning uses up a very definite amount of oxygen, we can also calculate the energy of the coal in calories by measuring the oxygen it consumes in burning. In the lamp, electric energy is converted into total radiant energy of all wave lengths, and this transformation may be very efficient in the case of a tungsten nitrogen filled lamp, about 81 per cent. But the most wasteful transformation comes when we consider the proportion of visible radiation in the total radiation. The visible radiation is only 3.6 per cent of the total radiation, and 96.4 per cent is waste heat, of no value in vision. The "over-all" efficiency is the product of all these efficiencies and represents the per cent of energy in the coal which appears as visible light. For the best incandescent lamp it is about one per cent. See Table I, p. 215.

Let us now apply this reasoning to luminous animals, regarded as power plants for illumination, and ask what fraction of the energy of the fuel (food) appears as light. No one has determined this for the firefly, and the investigation would present special difficulties because the firefly flashes, and flashing lights cannot be measured easily. Moreover, the firefly light is incidental to so many other life processes that its over-all efficiency would not be a fair comparison. It is undoubtedly very low. Concerning the radiational efficiency (energy in radiation divided by energy in chemiluminescent reaction) nothing is known. The luminous efficiency of the firefly we have already seen to be quite high.

Some organism, such as luminous bacteria whose light is continuous, would be best for efficiency studies. Such bacteria are fundamentally power plants for producing light and their luminous efficiency has been measured (p. 209). We cannot give the radiational efficiency since the heat of the chemiluminescent reaction in the bacterium is unknown, but the over-all efficiency can be easily determined.

We must measure the light produced by a single bacterium, and express this in calories per second. Just as we can measure the energy of coal by the oxygen it consumes in burning, so we can measure the energy of an animal's food, for instance sugar, from the total oxygen consumed in burning the sugar, and express the energy input in calories per second. Then, light emitted in calories divided by food oxidized in calories, gives the over-all efficiency of a bacterium.

The light measurements themselves present no particular difficulties. We can make an emulsion of luminous bacteria in sea water, many billions of them, count the number of bacteria per cubic centimeter, measure the amount of light emitted by one cubic centimeter, measure the absorption of light by bacteria in front of others, and calculate the amount of light, in lumens, which each bacterium would emit in all directions, provided there were no absorption. As one candle emits 4π lumens, the candle power of the smallest known light is easily obtained, about 20-quadrillionths of a candle for a single bacterium.

Metabolism experiments in animals are usually carried out by the method of "indirect calorimetry," by measuring oxygen consumed. A gram of tallow oxidized by a guinea pig liberates the same amount of heat during combustion to carbon dioxide and water as if it had been burned in a candle. This was one of Lavoisier's great contributions to science. A bacterium could obtain no more energy in burning its foodstuffs than a guinea pig or any other organism. Knowing the oxygen consumption of an animal and its food, we can calculate its heat production with results that are in surprising agreement with direct measurement of heat production in a calorimeter.

The author (1925) applied this method to luminous bacteria, whose oxygen consumption was measured when fed upon 60 per cent glycerine and 40 per cent peptone. With such a diet each milligram of oxygen consumed should produce 3.4 gram calories. Converting energy from lumens of light emitted into the same units—calories, the over-all efficiency of a bacterium turned out to be 0.16 per cent. This tells us the percentage which appears as light, of all the energy necessary to run a bacterium. It does not give us a true picture of the efficiency of the light-producing reaction, for much of the oxygen consumed is used by bacteria for growth processes which have nothing to do with luminescence.

To continue our comparison of a bacterium with a power plant, it is as if the power plant had to enlarge itself at the same time that it generated electricity. The power plant is a stable unit, and in comparing a bacterium and a power plant it is hardly fair to count the energy used in growth as part of that necessary for light production. Unfortunately we do not know what per cent of the oxygen consumed by a bacterium is used in growth or other processes but by treating bacteria with potassium cyanide, the total ogygen consumption can be reduced to one-twentieth of the previous value while the light is only slightly affected, perhaps reduced to one-quarter. These experiments indicate that at most only one-fifth of the oxygen is used in luminescence, and probably very much less than this. Allowing four-fifths of the oxygen for the non-luminescent oxygen consumption brings the efficiency of the bacterium to nearly 1 per cent (.16 per cent × 5), a figure about the same as that for over-all efficiency of the ordinary incandescent lamp calculated from the energy of coal used in generating current. Compared with chemiluminescences such as the oxidation of phosphorus (Adams, 1924), whose efficiency based on the heat of combustion of phosphorus is 0.017 per cent, or in Grignard compounds (Thomas and Dufford, 1933) of 0.0026 per cent, the efficiency is remarkably high.

By converting the light emitted into quanta for the maximum frequency in the bacterial light, we can calculate that one quantum of light appears for every 168 molecules of oxygen absorbed. If only 20 per cent of the oxygen is used in luminescence, 33 oxygen molecules entering the bacterium could produce one quantum. Eymers and van Schouwenburg (1937), in a similar investigation, found 193 molecules per quantum, based on 20 per cent of oxygen used in the light process. This efficiency varied with the temperature.

It would be interesting to compare the quantum efficiencies of bacteria with those of *Cypridina* and other chemiluminescences. A very rough measure of *Cypridina* luminescence made with crude extracts of the animal, indicated 50 molecules of oxygen consumed per quantum of $\lambda = 0.48\mu$ light emitted (Harvey, 1927), but this determination should be repeated with the purified material now available. Only one chemiluminescence has been investigated from the quantum viewpoint, that of luminol oxidized with hydrogen peroxide and hypochloride. In this case about 350 molecules are necessary to produce a quantum (Harris and Parker, 1935).

Table I gives some comparisons of the efficiencies of various illuminants, including a summary of what is known of the efficiency of animal light. It will be noted that for commercial illumination, total efficiencies are continually rising. Light

TABLE I. A COMPARISON OF THE EFFICIENCIES OF VARIOUS KINDS OF LAMPS AND LIGHTS

All percentages are based on 621 lumens per watt = 100% efficient. Values are from various sources; the sodium and fluorescent lamp data kindly supplied by Dr. C. G. Found of the General Electric Co.

Light	Commercial rating	A Coal to energy at lamp Food to bacterium %	B Radiational efficiency Radiant flux / Energy Input %	C Luminous efficiency Luminous flux / Radiant flux		Total luminous efficiency ($B \times C$) Luminous flux / Energy Input		Overall efficiency $A \times B \times C$	
				%	lumens/watt	%	lumens/watt	%	lumens/watt
Ideal	Yellow green, $\lambda = .555\mu$			100.0	621.0				
Black body	6500° K			13.9	86.3				
Sun				16.1	100.0				
Black body	3400° K			7.0	43.0				
Tungsten fil.	gas filled 30 kilowatt	33	90	5.5	34.4	5.0	31.0	1.6	10.3
"	gas " 200 watt	33	81	3.6	22.4	2.96	18.4	0.99	6.1
"	vacuum 40 watt	33	68	2.81	17.5	1.91	11.9	0.64	3.9
Carbon fil.	" 4 watts/candle	33	93	0.45	2.8	0.42	2.6	0.14	0.86
Welsbach	gas and mantle	80	16	1.2	7.5	0.19	1.2	0.15	0.96
Gas	open burner	80	19	0.19	1.2	0.036	0.22	0.03	0.18
Sodium	vapor NA9	33	55	16.4	102.0	9.0	55.5	3.0	18.5
Fluorescent	48″ white, 40 watt	30*	62.5	12.8	79.5	8.0	50.0	2.5	15.0
Firefly				80–92	493–571				
Cypridina				20.0	124.0				
Bacteria	(without cyanide)			45.0	279.0			0.16	1.0
"	(with cyanide)			45.0	279.0			0.80	5.0

* This value includes the loss in the lamp transformer.

from incandescent sources has been pushed to a temperature of 3670° C. Gas-filled bulbs and refractory material prevent evaporation of the heated filament. The maximum efficiency from such sources is from 20 to 30 lumens per watt input. The reader is referred to papers of Dushman (1934, 1937) and Found (1937), who have given an excellent analysis of the efficiencies and characteristics of both incandescent and gas discharge lamps.

Electroluminescences have long been recognized as highly efficient. They also use little current. A neon glow lamp for marking fire-escapes consumes only 2 watts. The color renders it impossible for ordinary lighting but for advertising and for signs, gas discharge tubes leave little to be desired.

Fortunately, the color can be corrected by combining the colors of several discharge tubes, or better, by combining the electroluminescence of the gas with the fluorescence of some added powder excited by the electroluminescent radiation, much of which is in the ultraviolet. In this way tubes emitting white light of low intrinsic brightness but of large area can be constructed with an efficiency of 50 lumens per watt input. These lamps are probably destined to replace the incandescent bulb. We can say that commercial lighting has already turned from incandescence to luminescence, thus adopting the method of the firefly.

INVISIBLE RADIATION

No book on animal luminescence would be complete without reference to invisible or penetrating radiation or emanation produced by living things or organic material. It is not without significance that immediately following the discovery of X-rays in 1895, Muraoka (1896) and later Singh and Maulik (1911) described radiations coming from fireflies which would pass opaque objects and affect a photographic plate. Dubois (1901) reported the same from luminous bacteria. The existence of such radiation has been denied by Suchsland (1898), Schurig (1901) and Molisch (1904). The experiments of Molisch are of greatest interest, for they are very carefully controlled and show without a doubt that black paper or metal sheet will allow no rays from these organisms to pass that will affect a photographic plate, even after several days' exposure. The *visible* light of luminous bacteria will affect the plate after one second exposure. Moreover, Molisch has pointed out the errors of those who claim to have found penetrating radiation in luminous forms. Certain kinds of cardboard, especially yellow varieties, or wood, will give off vapors that affect the photographic plate. The action is especially marked with damp cardboard at a temperature of 25° to 35° C., and Dubois and Muraoka must have used such cardboard to cover their plates. A piece of old dry section of beech or oak trunk, placed on a photographic plate for 15 hours in a totally dark place, will register a beautiful picture of the annual rings of growth, medullary rays, junction of bark and wood, etc. Moser (1842), Niepce (1857) and particularly Russell (1897-1908), had previously found that many bodies, both metals and substances of organic origin (resins, wood, paper, etc.), placed in contact with photographic plates, would affect them, and concluded that vapors and not rays were the active agents. As a dry piece of wood has a very definite smell, there is something given off which can affect the nose and there is no reason why it should not change, by purely chemical action, the photographic plate. This action of wood on the plate is prevented by interposing a sheet of glass. Frankland (1898) also noted vapors coming from the non-luminous species, *Bacillus proteus vulgaris*, and *B. coli communis*, which affected a photographic plate laid directly over the colonies in an open petri dish. There was no effect if the glass cover of the petri dish was between plate and bacteria.

This "Russell effect" is responsible for the fogging of plates by lecithin, blood and various organs of rabbits during oxidation (Schlaepfer, 1905) and for the apparent emission of ultraviolet light by cod liver oil and other antirachitic substances when oxidized (see Kugelmass and McQuarrie, 1924, 1925). Mathews and Dewey as well as Vincent and Morley (1913) also showed that the supposed radiation found by Matuschek and Nenning (1912) from chemical reactions, such as the solution of metals in acid, action of water on calcium carbide or formation of hydroxides, was in reality a chemical action of reducing or of other gases on the plate. Hydrogen peroxide, widely produced in many reactions, is extremely active and usually responsible for the effect. Keenan (1926) has reviewed the subject and literature, which is unusually extensive. Anyone can produce mysterious effects by placing a thin newly cleaned sheet of aluminum on a photographic plate in a dark box for 12 to 24 hours. When the plate is developed, a beautiful blackening will appear in the area covered by the aluminum. The formation of a film of aluminum hydroxide liberates hydrogen which forms hydrogen peroxide with oxygen of the air. As early as 1857 Niepce found that an engraving exposed to light and then placed on photographic paper would leave an impression.

We can attack the problem of penetrating radiation from luminous animals in another way. X-rays and radium rays (Becquerel rays) cause fluorescence of zinc sulphide, barium platinocyanide, willemite (Zn_2SiO_4), and calcium tungstate. Coblentz (1912) showed that the firefly will cause no fluorescence of a barium platinocyanide screen and I have been unable to detect fluorescence of zinc sulphide, barium platinocyanide, willemite or calcium tungstate shielded from *Cypridina* light by black paper, although the light of this organism is quite bright enough to cause phosphorescence of zinc sulphide without the black paper. The samples of the above four substances all showed fluorescence in presence of radium rays. There is, then, no specific emission of X-rays or of similar penetrating radiation from luminous tissues which will affect the photographic plate through opaque screens.

There is no reason why infrared or ultraviolet radiation should not be emitted during chemical reaction except that imposed by the energetics of the process. If we assume that each molecule which reacts emits a quantum of light, we can calculate the gram molecular heat equivalent of the quantum by multiplying a quantum of any frequency by the number of molecules in a gram molecular weight (6.06×10^{23}) and converting the ergs into calories. For visible frequencies ($\lambda = 0.40$ to 0.76μ) this turns out to vary from 71,000 to 37,000 calories. From any chemiluminescent reaction, the quantum energy of the radiation should not exceed the heat of reaction. Unless the heat of reaction is greater than 71,000 calories, we should not expect to find ultraviolet emitted and unless greater than 37,000 calories, no visible light. It is interesting that the heats of combustion of most chemiluminescent reactions do lie within this range and significant that 54,000 calories is a widely found value for heat of combustion per gram atom of oxygen for organic compounds and for oxidative dehydrogenations. This value corresponds to a calorie-quantum emission of green light, the commonest color in bioluminescent spectra.

However, too much weight should not be placed on the chemiluminescence equivalent, where the steps in the process are unknown. Actually a number of reactions have been studied where the heat of reaction is less than 71,000 calories and ultraviolet is produced. This is true for the thermal decomposition of ozone (Stuchtey, 1920; Wulf, 1926) and for combination of mercury and chlorine to form mercuric chloride (Haber and Zisch, 1922), both gas luminescences. It would seem possible, by excitation in successive stages, to build up molecules which could emit in the ultraviolet. No luminescent organism has been found to do so and inorganic chemiluminescences in the ultraviolet occur in the gas phase and are rare; although this may be in part due to the fact that they have not been systematically looked for. No intense ultraviolet has ever been observed in chemiluminescent reactions in aqueous solution.

One type of ultraviolet emission has received considerable attention in recent years, largely through the work of Audubert (1933-1939), made possible by development in the field of amplification and the perfection of the Geiger-Müller counter. This has been very aptly called a microluminescence, since only a few thousand quanta per cm.^{2} per second are emitted at $\lambda = 0.20\mu$. For $\lambda = 0.24\mu$, about three times as many quanta appear. Such reactions as the neutralization of strong acids with strong alkalies, oxidation at an aluminum anode, the oxidation of sodium pyrogallate or sulphite by oxygen, or of glucose by permanganate belong in this group. The radiochemical yield (ratio of quanta to molecules reacted) is of the order 10^{-14} to 10^{-15}. In the thermal decomposition of azides the emission is much greater and the radiochemical yield of the order 10^{-9} to 10^{-10}. Spectra have been obtained for decomposition of azides with maxima at $\lambda = 0.1975, 0.2150, 0.2300$ and 0.2400μ. It is perhaps too early to pass final judgment on these results but theoretical considerations indicate that such weak chemiluminescent reactions are possible (Evans, Eyring and Kincaid, 1938).

PHYSICAL NATURE OF LIGHT

MITOGENETIC RAYS

Of even less intensity are the mitogenetic or M-rays, short ultraviolet radiation, whose controversial history forms one of the most remarkable developments of recent years. Originally postulated to explain the initiation of mitotic cell division in injured tissues, the M-ray was at first believed to be an influence of unknown character but not a wound hormone. The M-rays passing out of the end of one root tip could start cell division in the near side of another root. The explanation of this basic experiment of Gurwitsch (1923, 1932, 1937) has so changed in the last fifteen years that the original idea of a dividing cell emitting something that would stimulate another cell to divide can hardly be recognized. The root tip as a detector has been almost completely replaced by microorganisms, particularly by the budding of yeast. Mitogenetic rays have been found emitted from sterile beet pulp, tadpole heads, yeast (but only in the light), bacteria, egg yolk, contracting muscle, active isolated nerve, frog heart, bone marrow, malignant tumors, normal and hemolysed blood, urine of healthy persons, etc., in many of which mitosis does not actively occur.

Later the rays were found to be produced during any kind of oxidative reaction and Gurwitsch (1925) has extracted from onion root pulp by the same procedure as for preparation of luciferin and luciferase, a "mitotin" and a "mitotase." Alone neither substance emits mitogenetic rays but mixed together, rays are produced for as long as one hour. Still later M-rays were found to be emitted *in vitro* by such processes as hydrolyses, neutralizations and the mere solution of sodium chloride in water.

The rays moved in straight lines but could be reflected, refracted and dispersed. Penetration through various screens (if we overlook the penetration necessary to emerge from the root tip) would indicate the rays had the character of ultraviolet light. Some workers found them to belong in the region, $\lambda = 0.19$ to 0.23μ, while others found a wave length of 0.33 to 0.34μ with a secondary maximum of 0.28μ. Truly astonishing are the mitogenetic spectra which have been published.

The biological method of detection is surest but physical methods have also been used. The photographic plate and particularly the photocell with the Geiger-Müller counter have led to divergent results in the hands of different investigators, chiefly because the intensity of the M-rays in the far ultraviolet is of the order of a few hundred quanta per cm.^{2} per second—an entirely new order of intensity and one which is at the limit of sensitivity of counters now available (Hollaender and Claus, 1937). This is millions of times smaller than the ultraviolet intensity which has been reported to stimulate cell division. Hollaender and Claus in a very carefully conducted investigation of mitogenetic rays failed entirely to find any stimulating effect of ultraviolet on yeast in the range 300 to 10^{12} quanta per cc. Only when the dose was increased to $50,000 \times 10^{12}$ quanta per cc. did effects appear that might possibly be interpreted as stimulation.

Most remarkable of all is the phenomenon of induced radiation—M-rays absorbed by cells and re-emitted as secondary radiation—so that a propagated mitogenetic impulse may spread throughout a tissue. This assumption might explain why very short ultraviolet light is not absorbed in passing longitudinally along a root tip but leads to other difficulties—why no lateral emission of M-rays appears from the root.

Indeed the whole story is so fantastic, the findings of different investigators are so contradictory, the views of the

Gurwitsch school have changed so suddenly, with no regard to what had previously been written, and statements of the properties of mitogenetic radiation are so at variance with what is known of the nature of ultraviolet light, that it is impossible to make logic out of the reports.

One always hesitates to deny the existence of a positive finding but it should have been possible to allay skepticism after eighteen years of research which has resulted in the publication of over 600 papers and several books.[6] We are greatly tempted to attribute mitogenetic rays to uncritical work with experimental material insufficiently controlled and primarily difficult to work with. At best, experimenters have been dealing with phenomena on the threshold of measurement, always a dangerous field for investigation. We are reminded of the famous N-rays of Blondlot (1904) and Charpentier (1904), also emitted by contracting muscles, conducting nerve and various tissues. Their detection depended on a change in brightness of a phosphorescent screen already near the threshold of vision. Their disproof came in part from the simple expedient of removing a prism essential to the demonstration but whose removal was unknown to those conducting the experiment. Lack of this vital part of the apparatus made no difference in the behavior of the N-rays and indicated that they were subjective in nature (Wood, 1904). A detailed account of the history and bibliography of this subject is given by Stradling (1907) in the *Journal of the Franklin Institute*.

[6] A short bibliography will be found under the heading "mitogenetic rays."

BIBLIOGRAPHY

BIBLIOGRAPHY

THE literature on luminescence is so vast that it is impossible to include in this book the important older papers, even though they are directly concerned with bioluminescence. Fortunately a number of bibliographies are available from which these references can be obtained. They have been marked with an asterisk and will be found in the section on general papers dealing with luminescence. Ehrenberg (1834) is especially good on early history.

Research on animal light has been an important field of study during the past twenty-five years at Princeton University. A list of publications from the Physiological Laboratory follows the section on general papers. Professor Ulric Dahlgren's comprehensive studies on histology will be found in the *Journal of the Franklin Institute* for 1915, 1916 and 1917.

The remaining sections contain only the literature from 1920 to date, classified under the various phyla of the animal kingdom and a few other headings. The author has endeavored to make the list fairly complete for the physiological and morphological (excluding systematic) fields as well as for chemiluminescences in solution. For phosphorescence, fluorescence and special subjects, the comprehensive monographs must be consulted. Kayser (1908), Becker (1923), Pringsheim (1928), Dankwortt (1929) and Radley and Grant (1933) contain many references, while Kayser (1908) gives a very complete history of our knowledge of non-living luminescences.

It is always a difficult task to include all publications in a field. No paper has been intentionally omitted and I hope few have been overlooked.

LIVING LIGHT
BOOKS AND GENERAL PAPERS ON BIOLUMINESCENCE
* Papers marked with an asterisk contain bibliographies.

BUCHNER, P. 1926. Tierisches Leuchten und Symbiose. Berlin. J. Springer. 58 pp.

———, 1930. Symbiose bei Leuchtenden Tieren. Tiere und Pflanze im Symbiose. Berlin. 2nd ed. pp. 664-736; 1st ed. 1921, pp. 340-400.

BÜTSCHLI, O. 1921. Vorlesung über vergleichende Anatomie. 1 Lfg. 3. Leuchtorgane. 6 Kap. 900-931. Berlin.

DAHLGREN, U. 1915-1917. Production of Light by Animals. Journ. Frank. Inst. 180, 513-537; 717-727; 181, 109-125; 243-261; 377-400; 525-556; 659-699; 805-843; 183, 79-94; 211-220; 323-348; 593-624; 735-754.

DITTRICH,* R. 1888. Über das Leuchten der Tiere. Wissensch. Beilage zum Program d. Realgymm. am Zwinger zu Breslau. pp. 1-70. Bibliography of 250 titles.

DUBOIS, R. 1914. La Vie et La Lumiere. Alcan. Paris. pp. 1-338.

EHRENBERG,* C. G. 1834. Das Leuchten des Meers. Abhandl. d. Kgl. Akad. d. Wiss. Berlin (1834), 411-572. History and references to 436 papers.

GADEAU DE KERVILLE, H. 1890. Les Vegetaux et les Animaux Lumineaux. Paris. 327 pp.

GIGLIONI, E. H. 1870. La Fosforescenza del Mare. Atti R. Accad. Sc. Torino, 5, 485-505.

HARVEY,* E. N. 1920. The Nature of Animal Light. J. B. Lippincott. Phila. 182 pp. 14 tables, 35 figs. Bibliography of 298 titles.

———,* 1924. Recent Advances in Bioluminescence. Physiol. Revs. 4, 639-671. Bibliography of 165 titles.

———, 1927. Bioluminescence. Bull. of Nat. Res. Council No. 59, 50-62.

———, 1935. Luciferase, the Enzyme Concerned in Luminescence of Living Organisms. Ergebnisse der Enzymforschung, 4, 365-379.

HEINRICH, P. 1815. Die Phosphorescenz der Körper. Das Leuchten der Pflanzen und Tieren. pp. 313-424. Nürnberg.

HELLER, J. 1853. Ueber das Leuchten im Pflanzen- und Tierreiche. Arch. f. physiol. u. pathol. Chem. u. Mikrosk. 6, 44, 81, 121, 161, 201, 241.

KLEIN, G. 1928. Die Lichtentwicklung bei Pflanzen. Handb. der Normalen und Pathologischen Physiologie 8, (2), 1057-1071.

KRUKENBERG, F. C. W. 1887. Neue Tatsachen für eine vergleichende Physiologie der Phosphorescenzerscheinungen bei Tieren und bei Pflanzen. Vergl. Physiol. Studien, II. Reihe, 4. Abt., Heidelberg, pp. 77-142.

MACARTNEY, J. 1810. Observations upon Luminous Animals. Phil. Trans. Roy. Soc. 100, 258-293.

BIBLIOGRAPHY

McDermott, F. Alex. 1911. Recent Advances in our Knowledge of the Production of Light in Living Organisms. Smithsonian Report, 345-362.

Mangold,* E. 1910. Die Produktion von Licht. Winterstein's Handbuch der vergleichende Physiologie 3, (2nd half) pp. 225-392. Jena. Bibliography of 649 titles.

———, 1925. Chemie der Lichtproduktion durch Organismen. Oppenheimer's Handb. d. Biochem. 2nd ed. 2, pp. 433-441.

———, 1928. Produktion von Lichtenergie bei Tieren. Bethe's Handb. der Normalen und Pathologischen Physiologie 8, (2) 1072-1082.

Molisch, H. 1904; 1912. Leuchtende Pflanzen. Eine Physiologische Studie. Jena. 1st ed. 1904, pp. 1-168. 2nd ed. 1912, pp. 1-198.

Panceri, M. 1872. Etudes sur la Phosphorescence des Animaux Marins. Ann. de Sc. Nat. 5 ser. 16, article 8, 66 pp.

Phipson, T. L. 1870. Phosphorescence. London. 210 pp.

Pierantoni, U. 1918. Les Microorganismes Physiologiques et la Luminescence des Animaux. Scientia 23, 43-53.

Pratje,* A. 1923. Das Leuchten der Organismen. Eine Übersicht über die neuere Literatur. Ergeb. d. Physiol. 21, 1-108. Bibliography of 501 titles.

Pütter, A. 1905. Leuchtende Organismen. Ztschr. f. allg. Physiol., Sammelreferate 5, 17-53.

Quatrefages, A. de. 1850. Mémoire sur la Phosphorescence de Quelques Invertébrés Marins. Ann. d. Sc. Nat. 3. ser., Zool. 14, 236-281; also Pop. Sc. Rev. 1, 275-298, 1862.

Spallanzani, L. 1784. Lettera Prima Relativa a Diversa Produzioni Marine. Mem. de Mat. e. Fis. d. Soc. Ital., 2, 603-661; on Meduse Fosforische, 7, 271-290, 1794.

Tilesius von Tilenau, W. G. 1819. Leuchten des Meers. Gilbert's Ann. d. Phys. 61, 36-44, 142-160, 161-176.

Trojan, E. 1917. Die Lichtentwicklung bei Tieren. Intern. Zeit. phys.-chem. Biol. 3, 94-105. Also Naturwiss. Wochenschrift 16, 457-462.

Viviani, D. 1805. Phosphorescencia maris. Quattuordecim Lucescentium Animaculorum Novis Speciebus Illustrata. Genoa. 17 pp. 5 plates.

Watasé, S. 1895. On the Physical Basis of Animal Phosphorescence. Biol. Lect. Marine Biol. Lab., Woods Hole, (1895) 101-119; also (1898) 177-192.

CONTRIBUTIONS FROM THE PHYSIOLOGICAL LABORATORY, PRINCETON UNIVERSITY

E. NEWTON HARVEY

1913. The Temperature Limits of Phosphorescence of Luminous Bacteria. Bioc. Bull., 2, 456-457.

1914. On the Chemical Nature of the Luminous Material of the Fire-fly. Science, 40, 33-34.

E. NEWTON HARVEY (CONTINUED)

1915. Experiments on the Nature of the Photogenic Substance in the Fire-fly. J. Am. Chem. Soc., 37, 396-401.
1915. Studies on Light Production by Luminous Bacteria. Am. J. Physiol., 37, 230-240.
1915. The Effect of Certain Organic and Inorganic Substances upon Light Production by Luminous Bacteria. Biol. Bull., 29, 308-312.
1916. The Mechanism of Light Production in Animals. Science, 44, 208-209.
1916. Studies on Bioluminescence. II. On the Presence of Luciferin in Luminous Bacteria. Am. J. Physiol., 41, 449-453.
1916. Studies on Bioluminescence. III. On the Production of Light by Certain Substances in the Presence of Oxidases. Am. J. Physiol., 41, 454-464.
1916. The Light Producing Substances, Photogenin and Photophelein, of Luminous Animals. Science, 44, 652-654.
1917. An Instance of Apparent Anesthesia of a Solution. Am. J. Physiol., 42, 606.
1917. The Chemistry of Light-Production in Luminous Organisms. Pub. Carnegie Inst. Wash. No. 251, 171-234.
1917. Studies on Bioluminescence. IV. The Chemistry of Light Production in a Japanese Ostracod Crustacean, Cypridina hilgendorfii, Müller. Am. J. Physiol., 42, 318-341.
V. The Chemistry of Light Production by the Fire-fly. Am. J. Physiol. 42, 342-348.
VI. Light Production in a Japanese Pennatulid, Cavernularia haberi. Am. J. Physiol., 42, 349-358.
1917. What Substance is the Source of Light in the Fire-fly? Science, 46, 241-243.
1917. Studies on Bioluminescence. VIII. The Mechanism of the Production of Light during Oxidation of Pyrogallol. J. Biol. Chem., 31, 311-336.
1918. Studies on Bioluminescence. VII. Reversibility of the Photogenic Reaction in Cypridina. J. Gen. Physiol., 1, 133-145.
1919. Studies on Bioluminescence. IX. Chemical Nature of Cypridina luciferin and Cypridina luciferase. J. Gen. Physiol., 1, 269-293.
1919. Studies on Bioluminescence. X. Carbon Dioxide Production during Luminescence of Cypridina luciferin. J. Gen. Physiol., 2, 133-136.
XI. Heat Production during Luminescence of Cypridina luciferin. J. Gen. Physiol., 2, 137-145.
1920. Studies on Bioluminescence. XII. The Action of Acid and of Light in the Reduction of Cypridina Oxyluciferin. J. Gen. Physiol., 2, 207-213.

BIBLIOGRAPHY

E. NEWTON HARVEY (CONTINUED)

1920. Is the Luminescence of Cypridina an Oxidation? Am. J. Physiol., 51, 580-587.
1920. Further Studies on the Chemistry of Light Production in Luminous Organisms. Pub. Carnegie Inst. Wash., No. 281, 75-110.
1921. Animal Light. Trans. Illum. Eng. Soc. 7, 319-330.
1921. A Fish with a Luminous Organ Designed for the Growth of Luminous Bacteria. Science 53, 314-315.
1921. Studies on Bioluminescence. XIII. Luminescence in the Coelenterates. Biol. Bull. 41, 280-287.
1922. Studies on Bioluminescence. XIV. The Specificity of Luciferin and Luciferase. J. Gen. Physiol. 4, 285-295.
1922. The Permeability of Cells for Oxygen and its Significance for the Theory of Stimulation, J. Gen. Physiol., 5, 215-222.
1922. Cold Light. Scribners' Mag., 72, 455-466. (October).
1923. Animal Luminescence. J. Frank. Inst., 189, 31-44.
1923. Studies on Bioluminescence. XV. Electroreduction of Oxyluciferin. J. Gen. Physiol., 5, 275-284.
1923. The Production of Light by the Fishes, Photoblepharon and Anamolops. Contributions from the Tortugas Laboratory. Pub. Carnegie Inst. Wash. No. 312, 43-60.
1923. The Minimum Concentration of Luciferin to give a Visible Luminescence. Science 57, 501-503.
1923. (and T. F. Morrison). The Minimum Concentration of Oxygen for Luminescence by Luminous Bacteria. J. Gen. Physiol., 6, 13-19.
1924. Studies on Bioluminescence. XVI. What Determines the Color of the Light of Luminous Animals? Am. J. Physiol., 70, 619-623.
1924. Recent Advances in Bioluminescence. Physiol. Rev. 4, 639-671.
1925. Studies on Bioluminescence. XVII. Fluorescence and Inhibition of Luminescence in Ctenophores by Ultraviolet Light. J. Gen. Physiol. 7, 331-339.
1925. The Inhibition of Cypridina Luminescence by Light. J. Gen. Physiol., 7, 679-685.
1925. The Effects of Light on Luminous Bacteria. J. Gen. Physiol., 7, 687-691.
1925. The Total Efficiency of Light Producing Organisms. News Letter of Princeton Eng. Assn., 117-119.
1925. Luminous Fishes of the Banda Sea. Natural Hist. 25, 353-356.
1925. The Total Efficiency of Luminous Bacteria. Jacques Loeb Memorial Volume of J. Gen. Physiol., 8, 89-108.
1926. On the Inhibition of Animal Luminescence by Light. Biol. Bull. 51, 85-88.
1926. Oxygen and Luminescence with a Description of Methods for Removing Oxygen from Cells and Fluids. Biol. Bull. 51, 89-97.
1926. Luminous Bacteria, the Smallest Lamps in the World. Scientific American, 414-416.

E. NEWTON HARVEY (CONTINUED)

1926. Additional Data on the Specificity of Luciferin and Luciferase, together with a General Survey of this Reaction. Am. J. Physiol., 77, 548-554.
1926. Bioluminescence and Fluorescence in the Living World. Am. J. Physiol., 77, 555-561.
1926. Further Studies on the Inhibition of Cypridina Luminescence by Light, with some Observations on Methylene Blue. J. Gen. Physiol., 10, 103-110.
1927. The Oxidation-Reduction Potential of the Luciferin-Oxyluciferin System. J. Gen. Physiol., 10, 385-395.
1927. On the Quanta of Light Produced and the Molecules of Oxygen Utilized during Cypridina Luminescence. J. Gen. Physiol., 10, 875-881.
1927. Luminous Animals. Scientia (May), 343-354.
1927. Bioluminescence. Bull. Nat. Res. Coun., No. 59, 50-62.
1927. Cold Light. Smithsonian Report for 1926, 209-218.
1928. Luciferase. Oppenheimer-Pincussen, Die Fermente und ihre Wirkungen. Bd. III: Die Methodik der Fermente, pp. 1400-1413.
1928. Luciferin and Luciferase, the Luminescent Substances of Light-giving Animals. Alexander's Colloid Chemistry, 2, 395-402.
1928. The Oxygen Consumption of Luminous Bacteria. J. Gen. Physiol., 11, 469-475.
1928. Photosynthesis in Absence of Oxygen. Plant Physiol. 3, 85-89.
1928. Studies on the Oxidation of Luciferin without Luciferase and the Mechanism of Bioluminescence. J. Biol. Chem. 78, 369-375.
1928. (and K. P. Stevens). The Brightness of the Light of the West Indian Elaterid Beetle, Pyrophorus. J. Gen. Physiol., 12, 269-272.
1928. Stability of Luminous Substances of Luminous Animals. Proc. Soc. Exp. Biol. and Med., 26, 133-134.
1929. (and R. T. Hall). Will the Adult Fire-fly Luminesce if Its Larval Organs are Entirely Removed? Science, 69, 253-254.
1929. (and A. L. Loomis). The Destruction of Luminous Bacteria by High Frequency Sound Waves. J. Bacteriol. 17, 373-376.
1929. A Preliminary Study of the Reducing Intensity of Luminous Bacteria. J. Gen. Physiol., 13, 13-20.
1929. Luminescence during Electrolysis. J. Phy. Chem. 33, 1456-1469.
1930. (and J. E. Deitrick). The Production of Antibodies for Cypridina Luciferase and Luciferin in the Body of a Rabbit. J. Immunology, 18, 65-71.
1930. (and P. A. Snell). The Kinetics of Bioluminescent Reactions of Short Duration. Proc. Am. Philos. Soc., 69, 303-308.
1931. (and P. A. Snell). The Analysis of Bioluminescences of Short Duration, Recorded with Photoelectric Cell and String Galvanometer. J. Gen. Physiol., 14, 529-545.

BIBLIOGRAPHY

E. NEWTON HARVEY (CONTINUED)

1930. Über Luciferase von Leuchtenden Tieren. Handb. d. biolog. Arbeitsmethoden, IV, (1), 827-853.
1931. Cold Light. The Scientific Monthly, 32, 270-272.
1931. (G. W. Taylor and). The Theory of Mitogenetic Radiation. Biol. Bull., 61, 280-293.
1931. (and G. I. Lavin). Reduction of Oxyluciferin by Atomic Hydrogen. Science, 74, 150.
1931. Photocell Analysis of the Light of the Cuban Elaterid Beetle, Pyrophorus. J. Gen. Physiol., 15, 139-145.
1931. Stimulation by Adrenalin of the Luminescence of Deep Sea Fish. Zoologica, 12, 67-70.
1931. Chemical Aspects of the Luminescence of Deep Sea Shrimp. Zoologica, 12, 70-75.
1932. The Evolution of Bioluminescence and Its Relation to Cell Respiration. Proc. Am. Philos. Soc., 71, 135-141.
1934. (and G. W. Taylor). The Oxygen Consumption of Luminous Bacteria in Water containing Deuterium Oxide. J. Cell. Comp. Physiol., 4, 357-362.
1934. (R. S. Anderson and). The Effect of Deuterium Oxide on the Luminescence of Luciferin. J. Cell. Comp. Physiol., 5, 249-253.
1935. Luciferase, the Enzyme Concerned in Luminescence of Living Organisms. Ergeb. der Enzymforschung, 4, 365-379.
1935. The Mechanism and Kinetics of Bioluminescent Reactions. Cold Spring Harbor Symposia on Quantitative Biology, III, 261-265.
1937. (F. H. Johnson and). The Osmotic and Surface Properties of Marine Luminous Bacteria. J. Cell. Comp. Physiol., 9, 363-380.
1938. (F. H. Johnson and). Bacterial Luminescence, Respiration and Viability in Relation to Osmotic Pressure and Specific Salts of Sea Water. J. Cell. Comp. Physiol., 11, 213-232.
1938. (and I. M. Korr). Luminescence in Absence of Oxygen in the Ctenophore, Mnemiopsis Leidyi. J. Cell. Comp. Physiol., 12, 319-323.
1939. The Luminescence of Adhesive Tape. Science, 89, 460-461.
1939. The Luminescence of Sugar Wafers. Science, 90, 35-36.
1939. Sonoluminescence and Sonic Chemiluminescence. J. Am. Chem. Soc., 61, 2392.
1939. Deep-sea Photography. Science, 90, 187.
1939. Bioluminescence. Trans. Farad. Soc. 35, 233-235.
1940. (with B. Chance, F. H. Johnson and G. Millikan). The Kinetics of Bioluminescent Flashes. A Study in Consecutive Reactions. J. Cell. Comp. Physiol., 15.
1940. (and R. S. Anderson). Luziferase. Die Methodik der Fermente, 2nd ed.

LIVING LIGHT

W. R. AMBERSON

1922. Kinetics of the Bioluminescent Reaction in Cypridina. I and II. J. Gen. Physiol., 4, 517-558.

R. S. ANDERSON

1933. The Chemistry of Bioluminescence. I. Quantitative Determination of Luciferin. J. Cell. Comp. Physiol., 3, 45-60.
1935. Studies in Bioluminescence. II. The Partial Purification of Cypridina Luciferin. J. Gen. Physiol., 19, 301-305.
1936. Chemical Studies on Bioluminescence. III. The Reversible Reaction of Cypridina Luciferin with Oxidizing Agents and Its Relation to the Luminescent Reactions. J. Cell. Comp. Physiol., 8, 261-276.
1937. Chemical Studies on Bioluminescence. IV. Salt Effects on the Total Light Emitted by a Chemiluminescent Reaction. J. Am. Chem. Soc., 59, 2115-2117.

A. M. CHASE

1940. Changes in the Absorption Spectrum of Cypridina Luciferin Solutions during Oxidation. J. Cell. Comp. Physiol., 15.
1940. Riboflavin and the Photochemical Oxidation of Cypridina Luciferin. Am. J. Physiol.

W. S. CREIGHTON

1926. The Effect of Adrenalin on the Luminescence of Fireflies. Science, 63, 600-601.

E. BROWNE HARVEY

1917. A Physiological Study of Specific Gravity and of Luminescence in Noctiluca, with Special Reference to Anesthesia. Pub. Carnegie Inst., Wash., No. 251, 235-253. Also Proc. Nat. Ac. Sc. 3, 15-16.

S. E. HILL

1928. The Influence of Molds on the Growth of Luminous Bacteria in Relation to the Hydrogen Ion Concentration, Together with the Development of a Satisfactory Culture Method. Biol. Bull., 55, 143-150.
1928. A Simple Visual Method for Demonstrating the Diffusion of Oxygen Through Rubber and Various Other Substances. Science 67, 374-376.
1929. (and C. S. Shoup). Observations on Luminous Bacteria. J. Bacteriology, 18, 95-99.
1929. The Penetration of Luminous Bacteria by the Ammonium Salts of the Lower Fatty Acids. Part I. General Outline of the Problem

BIBLIOGRAPHY

and the Effects of Strong Acids and Alkalies. J. Gen. Physiol, 12, 863-872.

1932. The Effects of Ammonia, of the Fatty Acids, and of their Salts, on the Luminescence of Bacillus Fischeri. J. Cell. Comp. Physiol., 1, 145-158.

F. H. JOHNSON

1935. Oxidation of Carbohydrates and Polyhydric Alcohols by Luminous Bacteria. Proc. Soc. Exp. Biol. and Med., 32, 1263-1265.

1935. A Micro-method for Determining the Utilization of Carbohydrates and Polyhydric Alcohols by Microorganisms. Science, 81, 620-621.

1936. The Aerobic Oxidation of Carbohydrates by Luminous Bacteria and the Inhibition of Oxidation by Certain Sugars. J. Cell. Comp. Physiol., 8, 439-463.

1936. (and I. V. Shunk). An Interesting New Species of Luminous Bacteria. J. Bacteriology, 31, 585-592.

1937. Hexose Oxidation by Luminous Bacteria. I. The Effect of Some Natural Synthetic Glycosides and Related Substances. J. Cell. Comp. Physiol., 9, 199-206.

1937. An Improved Thunberg Technique for Bacterial Oxidations. Proc. Soc. Exp. Biol. and Med., 36, 387-390.

1938. (and R. S. Anderson). Hexose Oxidation in Luminous Bacteria. II. The Inhibition of Glucose Oxidation by Alpha Methylglucoside. J. Cell. Comp. Physiol., 12, 273-281.

1938. Hexose Oxidation in Luminous Bacteria. III. The Escape of Respiration and Luminescence from Inhibition by Methylglucoside, with a Note on Urethanes. J. Cell. Comp. Physiol., 12, 281-294.

1939. (and E. L. Chambers). Oxygen Consumption and Methylene Blue Reduction in Relation to Barbital Inhibition of Bacterial Luminescence. J. Cell. Comp. Physiol., 13, 263-267.

1939. Decomposition of Hydrogen Peroxide by Catalase. Nature, 144, 634.

1939. Total Luminescence of Bacterial Suspensions in Relation to the Reactions Concerned in Luminescence. Enzymologia, 7, 72-81.

1939. (and K. L. van Schouwenburg and A. van der Burg). The Flash of Luminescence Following Anaerobiosis of Luminous Bacteria. Enzymologia, 7, 195-224.

I. M. KORR

1935. The Relation Between Cell Integrity and Bacterial Luminescence. Biol. Bull., 68, 347-354.

1935. An Electrometric Study of the Reducing Intensity of Luminous Bacteria in the Presence of Agents Affecting Oxidations. J. Cell. Comp. Physiol., 6, 181-216.
1936. The Luciferin-oxyluciferin System. J. Am. Chem. Soc., 58, 1060.

T. F. MORRISON

1925. II. The Influence of Temperature on the Intensity of the Light of Luminous Bacteria. J. Gen. Physiol., 7, 741-753.
1925. The Effect of Polarized Light on the Growth of Luminous Bacteria. Science, 69, 392-393.

C. W. ROOT

1932. The Relation Between Respiration and Light Intensity of Luminous Bacteria, with Special Reference to Temperature. I. Temperature and Light Intensity. J. Cell. Comp. Physiol., 1, 195-208.
1934. II. Temperature and Oxygen Consumption. J. Cell. Comp. Physiol., 5, 219-229.

H. SHAPIRO

1934. The Light Intensity of Luminous Bacteria as a Function of Oxygen Pressure. J. Cell. Comp. Physiol., 4, 313-328.

C. S. SHOUP

1929. The Respiration of Luminous Bacteria and the Effect of Oxygen Tension upon Oxygen Consumption. J. Gen. Physiol., 13, 27-45.
1930. The Luminous Bacteria. J. Tenn. Acad. Sci., 5, 1-4.
1928. Preservation of Luminous Bacteria in Absence of Oxygen. Proc. Soc. Exp. Biol. and Med., 25, 570-572.
1934. (and A. Kimler). The Sensitivity of the Respiration of Luminous Bacteria for 2, 4-dinitrophenol. J. Cell. Comp. Physiol., 5, 269-276.

P. A. SNELL

1931. The Neuro-muscular Mechanism Controlling Flashing in the Lampyrid Fireflies. Science, 73, 372-373.
1932. The Control of Luminescence in the Male Lampyrid Firefly, Photuris pennsylvanica, with Special Reference to the Effect of Oxygen Tension on Flashing. J. Cell. Comp. Physiol., 1, 37-51.

K. P. STEVENS

1927. Studies on the Amount of Light Emitted by Mixtures of Cypridina Luciferin and Luciferase. J. Gen. Physiol., 10, 859-873.

G. W. TAYLOR

1932. The Effects of Hormones and Certain Other Substances on Cell (Luminous Bacteria) Respiration. J. Cell. Comp. Physiol., 1, 297-331.

BIBLIOGRAPHY

1932. (O. W. Richards and). Mitogenetic Rays—A critique of the Yeast Detector Method. Biol. Bull., 63, 113-128.
1934. The Effect of Narcotics on Respiration and Luminescence in Bacteria with Special Reference to the Relation between the Two Processes. J. Cell. Comp. Physiol., 4, 329-356.
1936. The Effect of Ethyl Urethane on Bacterial Respiration and Luminescence. J. Cell. Comp. Physiol., 7, 409-416.

LUMINOUS BACTERIA

CLAREN, O. B. 1938. Zum Stoffwechsel der Leuchtbakterien. I. Ann. d. Chemie, 535, 122-149.

DeLERMA, B. 1937. Le Recenti Ricerche Chimico-fisiche sulla Bioluminescenzo (cm d'analisi spettrale della luce emessa do Bacillus sepiae Zirp). Riv. Fis. Mat. Sci. Nat., 11, 615-622.

DOUDOROFF, M. 1938. Lactoflavin and Bacterial Luminescence. Enzymologia, 5, 239-243.

EGOROWA, A. A. 1929. Leuchtbakterien im Schwarzen und im Asowschen Meere. Zeit. Bakter. II 79, 168-173.

EYMERS, J. G. and K. L. VAN SCHOUWENBURG. 1936. On the Luminescence of Bacteria. I. A quantitative study of the spectrum of the light emitted by Photobacterium phosphoreum and by some chemiluminescent reactions. Enzymologia, 1, 107-119.

———, 1937. On the Luminescence of Bacteria. II. Determination of the oxygen consumed in the light emitting process of Photobacterium phosphoreum. Enzymologia, 1, 328-340.

———, 1937. On the Luminescence of Bacteria. III. Further quantitative data regarding spectra connected with bioluminescence. Enzymologia, 3, 235-241.

FEJGIN, B. 1926. Études sur les Microbes Marins. Etude sur une Bacterie Lumineuse. Bull. Inst. Oceanographique, No. 471, 1-4.

FUHRMANN, F. 1932. Studien zur Biochemie der Leuchtbacterie. I. Der Einfluss von Na und K Chlorid und Bromid auf die Lichtentwicklung von Photobacillus radians. II. Der Einfluss von Zuckern mit NaCl. Monatsh. f. Chem., 60, 69-105; 414-430.

GERRETSEN, F. C. 1920. Über die Ursachen des Leuchtens der Leuchtbakterien. Zentralb. f. Bact. 2 Abt., 52, 353-373.

HARVEY, E. N. 1913, 1915, 1916, 1923, 1925, 1928, 1929, 1932, 1934, 1937, 1938.

HILL, S. E. 1928, 1929, 1932.

HSU, H. L. 1937. Ueber den Einfluss des Ionisierten Luftbades auf die Bakterien. Sei-i-kai Med. J., 56, 1-11.

INMAN, O. L. 1927. A Pathogenic Luminous Bacterium. Biol. Bull., 53, 197-200.

JOHNSON, F. H. 1935, 1936, 1937, 1938, 1939.

JOHNSON, F. H. and E. N. HARVEY, 1937, 1938.

Kishitani, T. 1933. Zur Morphologie und Biologie einer Leuchtbakterienart. J. Sc. Hiroshima Univ. B, 1, 183-196.
Kishitani, T. 1928. Drei Neue Arten von Leuchtbakterien. Proc. Imp. Acad. Tokyo, 4, 69-75.
Korr, I. M. 1935.
Kostka, G. 1928. Lebende Bakterien als Sauerstoff-indikatoren. Mikrokosmos, 22, 6-11.
Lebenbaum, M. 1930. L'influence des Ions sur la Luminescence Bacterienne. Acta Soc. Botanicorum Poloniae 7, 583-597.
Majima, R. 1931. Studies on Luminous Bacteria. Sei-i-Kwai Med. J., 50. I. Further Studies on the Pathogenic Vibrio, Microspira Phosphoreum of Yasaki, 1-23. II. Photogenic Bacteria Found in "Yanagimushi-Karei," 33-54. III. On the Fermentation of Carbohydrate and the Immune Agglutination by Various Luminous Microorganisms, 41-67.
Meissner, G. 1926. Bakteriologische Untersuchungen uber die Symbiontischen Leuchtbakterien von Sepien aus dem Golf von Neapel. Centralb. Bakt., 67, 194-238. Biol. Zentralb., 46, 527-542.
Molisch, H. 1925. Botanische Beobachtungen in Japan. III. Uber das Leuchten des Schlachtviehfleisches in Sendai, Japan. Science Reports of the Tôhoku Imperial Univ., Sendai, Japan. 4th Ser., Biol., 1, 97-103.
Morrison, T. F. 1925.
Ninomiya, R. 1924. Der Einfluss von Antikörpern und Komplement auf Biologische Functionen von Bakterien. I. Der Einfluss specifischer Amboceptoren mit und ohne Komplementzusatz auf das Leuchtvermögen von Leuchtbakterien. Zeit. f. Immun. u. exp. Therap., 39, 498-512.
Pierantoni, U. 1920. Sul Significato Fisiologico della Simbiosi Eredetaria. Boll. d. Soc. Natur. Napoli, 33, 55-66.
———, 1923. I Recenti Ricerche sulla Simbiosi Fisiologica Ereditaria. Arch. di. Sci. Biol., 4, 229-237. Also, Atti d. Soc. Ital. per il Prog. d. Scienze, 13, 1-20, 1924.
Richter, O. 1926. Bakterienleuchten "ohne Sauerstoff." Zeit. wiss. Biol. Abt. E. Planta, 2, 569-587.
———, 1928. Natrium ein Notwendiges Nährelement für eine Marine Microärophile Leuchtbacterie. Akad. wiss. Wien. Math. Nat. Denksch., 101, 261-292.
Root, C. W. 1932, 1934.
Shapiro, H. 1934.
Shoup, C. S. 1928, 1929, 1930, 1933, 1934.
Sonnenschein, C. 1931. Leuchtvibrionen als Sauerstoffzehrer bei der Anaerobenzuchtung und als Sauerstoffindikator. Zentb. f. Bak., Parasit. und Infeckt., 123, 378-381.
———, 1931. Fortzuchtung von Leuchtvibrionen in Rindergalle. Zentral. f. Bakt., Parasit. und Infekt., 123, 92-93.
———, 1932. Auf Leuchtvibrionen wirksame Bakteriophage. Centralb. f. Bak., 1st Ab., 126, 297-302.

BIBLIOGRAPHY

TAKASE, M. 1939. Untersuchungen über den Einfluss von vershiedenen Salzen auf die Lichtproduktion der Leuchtbakterien. Sei-i-kai-zassi, 58, 54-80.

TAYLOR, G. W. 1932, 1934, 1936.

VAN SCHOUWENBURG, K. L. 1938. On Respiration and Light Emission in Luminous Bacteria. Thesis. 97 pp. Delft, Holland.

VAN SCHOUWENBURG, K. L. and J. G. EYMERS. 1936. Quantum Relationship of the Light-emitting Process of Luminous Bacteria. Nature, 138, 245.

VOUK, V., V. SKORIC and Z. KLAS. 1931. A New Phosphorescent Bacterium from the Adriatic Sea and the pH Range of Its Luminosity. Bull. Intern. Acad. Yougoslave Sci. et Beaux Arts, 25, 86-88.

YASAKI, Y., M. NISHIO, A. ICHIKAWA, R. MAJIMA and O. ISHIKAWA. 1926. Bacteriological Studies on Bioluminescence. II. On the nature of the new luminous bacteria, Microspira phosphoreum Yasaki. Sei-i-kai-zasshi, 45.

ZIRPOLO, G. Studi sulla Bioluminescenza Batterica.

1920. I. Azione degli ipnotici. Riv. de Biol., 2, 52-59.

1920. II. Azione dei sali magnesia. Boll. Soc. Nat. Napoli, 32, 112-119.

1920. III. Azione dei raggi emanati dal bromuro di radio. Boll. Soc. Nat. Napoli, 33, 75-81.

1921. IV. Azione dei sali radioacttivi. Riv. di Sci. Nat., 12, 139-144.

1922. V. Azione del nitrato di cerio. Boll. Soc. Nat. Napoli, 34, 46-50.

1922. VI. Azione dei sali di chinina, coffeina, cocaina e strichnina. Rev. d Scienza Nat. Napoli, 13, 1-11.

1923. VII. Azione dei sali di potassio. Boll. Soc. Nat. Napoli, 35, 245-247.

1927. VIII. La resistenza del potere luminozo. Boll. Soc. Nat. Napoli, 18, 225-231.

1929. IX. Azione delle alte e basse temperature sui batteri luminosi. Boll. Soc. Nat. Napoli, 41, 137-149.

1931. X. Azione dei batteri luminosi sulla germenazione dei semi. Boll. Soc. Nat. Napoli, 43, 393-422.

1932. XI. Batteri luminosi ed. "Anelli di Lieseganz." Boll. Soc. Nat. Napoli, 44, 221-228.

1932. XII. Azione dell'idrogeno ($-253°C$) e dell'elio liquido ($-269°C$). Boll. Soc. Nat. Napoli, 44, 229-235.

1938. XIII. Azione dell'acqua pesante (D_2O). Boll. Zool., 9, 49-55.

———, 1926. Ancora sui Batteri Luminosi. Riv. d. Biol., 8, 244-248.

———, 1933. Ricerche Criobiologiche sui Batteri Luminosi dei Cefalopodi. Arch. Zool. Ital., 18, 359-405.

LIVING LIGHT

LUMINOUS FUNGI

BOSE, S. R. 1926. Luminous Leaves and Stalks from Bengal. Nature, 36, Jan. 30.

———, 1930. Relation of Sunlight to the Light of Luminous Wood. Naturwissenschaften, 36, 787.

———, 1935. A Luminous Agaric (Pleurotus sp.) from South Burma. Trans. Brit. Myc. Soc., 19, 97-101.

BOTHE, F. 1928. Über den Einfluss des Substrats und einiger anderer Faktoren auf Leuchten und Wachstum von Mycelium X und Agaricus melleus. Akad. wiss. Wien Math.-Natur. Kl. Sitzb. Abt. I, 137, 595-626.

———, 1930. Eine neuer Einheimisher Leuchtpilz. Ber. d.d. Bot. Ges., 48, 394-399.

———, 1931. Über das Leuchten verwesender Blätter und seine Erreger. Zeit. wiss. Biol. abt. E. Planta, 14, 752-765.

———, 1935. Genetische Untersuchungen über die Lichtentwicklung der Hutpilze. I. Arch. Protistenkunde, 85, 369-383.

BULLER, A. H. R. 1924. Researches on Fungi. Vol. III. The Bioluminescence of Panus stipticus. Chap. 12, 357-431. Longmans Green, London.

———, 1934. Vol. VI. Omphalia flavida, a geminiferous and luminous leaf-spot fungus. Chap. 3, 397-454.

HANNA, W. F. 1938. Notes on Clitocybe illudens. Mycologia, 30, 379-384.

MACRAE, R. 1937. Interfertility Phenomena of the American and European Forms of Panus stypticus (Bull.) Fries. Nature, 139, 674.

PORIFERA AND PROTOZOA

CARDOT, H. and M. LEFEVRE. 1929. Sur la Fonction Photogenique de Certains Peridiniens. Bull. Soc. Linn. d Lyons, 8, 48-50.

HARVEY, E. B. 1917.

KOFOID, C. A. 1920. A New Morphological Interpretation of the Structure of Noctiluca and its Bearing on the Status of Cystoflagellata (Haeckel). Univ. Cal. Pub. Zool., 19, 317-334.

KOFOID, C. A. and O. SWEZY. 1921. The Free-swimming Unarmoured Dinoflagellata. Mem. Univ. of Calif., 5, 1-562. 388 fig. in text. 12 plates.

OKADA, Y. K. 1925. Luminescence in Sponges. Science, 62, 567.

PRATJE, A. 1921. Die Verwandtschaftlichen Beziehungen der Cystoflagellaten zu den Dinoflagellaten. Arch. protistenk., 42, 422-437.

———, 1921. Noctiluca Miliaris, Beiträge zur Morphologie, Physiologie u. Cytologie. I. Morphologie u. Physiologie. Arch. f. Protistenkunde, 42, 1-95.

BIBLIOGRAPHY

PRATJE, A. 1921. II. Zur Chemie der Noctiluca Zellkernes. Zeit. f. Anat. u. Entwges., 62, 171-232.
———, 1921. Macrochemische, quantitative Bestimmung des Fettes u. Cholesterins, sowie ihrer Kennzahlen bei Noctiluca miliaris, sur. Biol. Zentb., 41, 433-446.
TROJAN, E. 1933. Light-producing Power of Sponges. Nature, 131, 728.

COELENTERATA AND CTENOPHORA

HARVEY, E. N. 1917, 1921, 1925, 1926, 1938.
HEYMANS, C. and A. R. MOORE. 1924. Luminescence in Pelagia noctiluca. J. Gen. Physiol., 6, 273-280.
———, 1925. Note on the Excitation and Inhibition of Luminescence in Beroë. J. Gen. Physiol., 7, 345-348.
HYKES, O. V. 1928. Contribution a la Physiologie de la Luminescence et de la Motilite des Coelenteres. C. R. Soc. Biol., 98, 259-261.
MOORE, A. R. 1923. Luminescence in Mnemiopsis. J. Gen. Physiol., 6, 403-412.
———, 1926. Galvanic Stimulation of Luminescence in Pelagia noctiluca. J. Gen. Physiol., 9, 375-381.
———, 1926. On the Nature of Inhibition in Pennatula. Am. J. Physiol., 76, 112-115.
———, 1926. The Photolysis of the Luminescent Granules of Eucharis multicornis. J. Gen. Physiol., 8, 303-310.
OKADA, Y. K. 1926. Light Localization in Ctenophores. Science, 63, 262.
PARKER, G. H. 1920. The Phosphorescence of Renilla. Proc. Am. Philos. Soc., 19, 171-175.
———, 1920. Activities of Colonial Animals. II. Neuromuscular movements and phosphorescence of Renilla. J. Exp. Zool., 31, 475-513.
PIERANTONI, U. 1924. II. La Fosforenza dei Ctenophori. Rend. R. Ac. Nat. d. Lincei, ser. 5, 33.

ANNELIDA AND NEMERTEA

FUJIWARA, T. 1935. On the Light Production and the Luminous Organs in a Japanese Chaetopterid, Mesochaetopterus japonicus, Fujiwara. J. of Sc. Hiroshima Univ., 3, 185-192.
GATES, G. E. 1925. Note on Luminescence in the Earthworms of Rangoon. Rec. of Indian Mus., 27, 471-473.
GILCHRIST, J. D. F. 1919. Luminosity and Its Origin in a South African Earthworm (Chitota sp.). Trans. Roy. Soc. S. Africa, 7, 203-212.
KATO, K. 1939. A New Luminous Species of the Nemertea, Emplectonema kandai sp. nov. Japanese J. of Zool., 8, 251-254.
KANDA, S. 1939. The Luminescence of a Nemertean, Emplectonema Kandai, Kato. Biol. Bull., 77, 166-173.
KOMAREK, J. 1934. Luminescence of Carpathian Worms and Its Causes. Bull. Internat. de l'Acad. Sc. de Boheme, 44, 1-10.

Komarek, J. 1936. Animalischen Leuchten. Biol. Listy, 21, 212-217.
Krekel, A. 1920. Die Leuchtorgane von Chaetopterus variopedatus, Clap. Zeit. wiss. Zool., 118, 480-509.
Meyer, A. 1929. Tomopteris Anadyomene nov. sp., ein Nachweis Phylogenetische Umwandlung von Nephridialtrichtern in Leuchtorgane bei den Polychaeten. Zool. Anz., 86, 124-133.
Pierantoni, U. 1924. La Fosforescenza e la Simbiosi in Microscolex phosphorens. Boll. d. Soc. d. Nat. Napoli, 36, 179-195.
———, 1923. Nuove Osservazioni su Luminescenzo e simbiosi. I. La foscorenza degli Oligocheti. Rend. R. Acc. Nat. d. Lincei, ser. 5, 32.
Skowron, S. 1926. On the Luminescence of Microscolex Phosphoreus Dug. Biol. Bull., 51, 199-208.
———, 1928. The Luminous Material of Microscolex Phosphoreus Dug. Biol. Bull., 54, 191-195.

CRUSTACEA

Amberson, W. R. 1922.
Anderson, R. S. 1933, 1935, 1936, 1937.
Anderson, R. S. and E. N. Harvey. 1934.
Chance, B., E. N. Harvey, F. H. Johnson and G. Millikan. 1940. The Kinetics of Bioluminescent Flashes. A Study in Consecutive Reactions. J. Cell. Comp. Physiol., 15.
Chase, A. M. 1940.
Coblentz, W. W. and C. W. Hughes. 1926. Spectral Energy Distribution of the Light Emitted by Plants and Animals. Sc. Papers Bur. of Standards, 21, 521-524.
Harvey, E. N. 1916, 1917, 1918, 1919, 1920, 1922, 1923, 1925, 1926, 1927, 1928, 1930, 1931, 1934, 1940.
Kanda, S. Physico-Chemical Studies on Bioluminescence.
  1920. I. On the luciferin and luciferase of Cypridina Hilgendorfii. Am. J. Physiol., 50, 544-560.
  1920. II. The production of light by Cypridina Hilgendorfii is not an oxidation. Am. J. Physiol., 50, 561-573.
  1921. IV. The physical and chemical nature of the luciferase of Cypridina Hilgendofii. Am. J. Physiol., 55, 1-12.
  1924. V. The physical and chemical nature of the Luciferine of Cypridina Hilgendorfii. Am. J. Physiol., 68, 435-444.
  1928. VI. The mechanism of luminescence in the Cypridina luciferin and luciferase suggested. Sci. Papers Inst. Phys. and Chem. Research, 9, 265-269.
  1929. VII. The solubility of Cypridina luciferin in organic solvents. Sci. Papers Inst. Phys. and Chem. Research, 10, 91-98.
———, 1930. The Chemical Nature of Cypridina luciferin. Science, 71, 444.
———, 1932. Crystalline Luciferin. Suppl. Sci. Papers Inst. Phys. and Chem. Research, 18, 1.

BIBLIOGRAPHY

KORR, I. M. 1936.
LEAVITT, B. B. 1935. A Quantitative Study of the Vertical Distribution of the Larger Zooplankton in Deep Water. Biol. Bull., 68, 115-130; 74, 376-394, 1938.
NICHOLS, E. S. 1922. A Convenient Method of Determining the Brightness of Luminescence. Science, 55, 157; see also 60, 592-593.
OKADA, YO K. 1927. Luminescence et Organe Photogené des Ostracodes. Bull. Soc. Zool., France, 51, 478-486.
———, 1928. Note on the Tail Organs of Ascetes. Ann. and Mag. Nat. Hist., ser. 10, 1, 308.
SHAPIRO, H. 1934.
STEUER, A. 1928. Über das sog. Leuchtorgan des Tiefsee-Copepoden, Cephalophanes, G. O. Sars. Arb. Zool. Inst. Univ. Innsbruck, 3, 9-16.
STEVENS, K. P. 1927.
TAKAGI, S. 1936. Über Sekretbildung in dem Leuchtorgan von Cypridina hilgendorfii, Müller mit besonderer Berücksichtigung der Mitochondrien. Annot. Zool. Japan., 15, 344-349.
WATERMAN, T. H., R. F. NUNNEMACHER, F. A. CHACE and G. L. CLARK. 1939. Diurnal Migration of Deep-water Plankton. Biol. Bull., 76, 256-279.
WELSH, J. H. and F. A. CHACE, JR. 1937. Eyes of Deep Sea Crustaceans. I. Acanthephyridae. Biol. Bull., 72, 57-74. II. Sergestidae, 74, 364-375.
WELSH, J. H., F. A. CHACE, JR. and R. F. NUNNEMACHER. 1937. The Diurnal Migration of Deep-water Animals. Biol. Bull., 73, 185-196.
YASAKI, Y. 1927. Bacteriological Studies on Bioluminescence. I. On the cause of luminescence in fresh water shrimp, Xiphocaridina compressa. J. Infec. Diseases, 40, 404-407.

MYRIAPODA, ARACHNIDA AND LOWER INSECTA

ARNDT, W. 1924. Leuchtende Tausendfüsse in Schlesein. J. H. d. Ver. schles. Insectenkunde, 14.
BEHNING, A. 1929. Über eine Leuchtende Chironomide des Tschalkar-Sees. Zeit. wiss. Insektenbiol., 24, 62-65.
BRADE-BIRKS, H. K. and S. G. 1920. Luminous Chilopoda. Hartford Naturalists' Field Club Occasional Papers, No. 20; also in Ann. Mag. Nat. Hist. ser. 9, 5, 1920. 30 pp.
BROWN, B. 1925. A Luminous Spider. Science, 62, 182, 329; 63, 383.
HEIDT, K. 1936. Über das Leuchten der Collembolen Onychiurus Armatus Tbg. und Achorutes Muscorum Templ. Biol. Zentralb., 56, 100-109.
HEPP, A. 1927. Biologische Beobachtungen (Grossechmetterlinge) Lepid. Rundschau, 1, 97-98.
HUDSON, G. V. 1926. "The New Zealand Glow-worm," Boletophila (Arachnocampa) luminosa: Summary of Observations. Ann. and Mag. Nat. Hist., 17, 228-235; 18, 667-670.

LIVING LIGHT

Isaak, J. 1916. Ein Fall von Leuchtfähigkeit bei einem Europaischen Grossmetterling. Biol. Zentralb., 36, 216-218; 37, 106-108.

Koch, A. 1927. Studien an Leuchtenden Tieren. I. Das Leuchten der Myriapoden. Zeit. wiss. Biol. Abt. A. Morph. u. Ökol. Tiere, 8, 241-270.

Pfeiffer, H. and H. J. Stammer. 1930. Pathogenes Leuchten bei Insekten. Zeit. f. Morph. und Ökol. der Tiere, 20, 136-171.

Stammer, H. J. 1932. Zur Biologie und Anatomie der Leuchtenden Pilzmückenlarve von Ceroplatus testaceus. Z. Morph. u. Ökol. Tiere, 26, 135-146.

―――, 1935. Das Leuchten des Collembolen, Achorutes Muscorum Templ., nebst Bemerkungen über die in Deutschland Vorkommenden Leuchtenden Landtiere. Biol. Zentralb., 55, 178-182.

―――, 1930. Das Gelegentliche Leuchten der Insekten Hervorgerufen. durch Pathogene Leuchtbakterien. Mitt. Deut. Entom. Ges., 1, 38-41.

COLEOPTERA (FIREFLIES)

Alexander, G. 1935. Is a Pacemaker Involved in Synchronous Flashing of Fireflies? Science, 82, 440.

Blair, K. G. 1927. An Aquatic Glowworm. Nat. Hist. Mag., 1, 59.

Brown, D., E. S. and C. V. King. 1931. The Nature of the Photogenic Response of Photuris pennsylvanica. Phys. Zool., 4, 287-293.

Buck, J. B. 1938. Synchronous Rhythmic Flashing of Fireflies. Quart. Rev. Biol., 13, 301-314.

―――, 1935. Synchronous Flashing of Fireflies Experimentally Produced. Science, 81, 339.

―――, 1937. Studies on the Firefly. I. The effects of light and other agents on flashing in Photinus pyralis with special reference to periodicity and diurnal rhythm. Physiol. Zool., 10, 45-58.

1937. II. The signal system and color vision in Photinus pyralis. Physiol. Zool., 10, 412-419.

―――, 1937. Flashing of fireflies in Jamaica. Nature, 139, 801.

Creighton, W. S. 1926.

Emerson, G. A. 1935. Some Effects of Ether on Bioluminescence in the Lampyrid, Photuris pennsylvanica. Proc. Soc. Exp. Biol. and Med., 33, 36-40.

Gerretsen, F. C. 1922. Einige Notizen über der Leuchten des Javanischen Leuchtkäfers (Luciola Vittata Cast.). Biol. Zentb., 42, 1-9.

Harvey, E. N. 1914, 1915, 1917, 1924, 1928, 1929, 1931.

Hess, W. N. 1920. Notes on the Biology of Some Common Lampyridae. Biol. Bull., 38, 39-76.

―――, 1922. Origin and Development of the Light-Organs of Photuris pennsylvanica. De Geer. J. Morph., 36, 244-277.

―――, 1921. Tracheation of the Light-Organs of Some Common Lampyridae. Anat. Rec., 20, 155-161.

BIBLIOGRAPHY

IVES, H. E. 1922. The Firefly as an Illuminant. J. Frank. Inst., 194, 213-230.

KANDA, S. 1920. Physico-Chemical Studies on Bioluminescence. III. The production of light by Luciola viticollis is an oxidation. Am. J. Physiol., 53, 137-149.

KNIPP, C. T. 1939. Discussion on the Path of the Firefly while Periodically Flashing. Science, 89, 386-387.

MALUF, N. S. R. 1938. The Basis of the Rhythmic Flashing of the Firefly. Ann. Entom. Soc. Amer. 31, 374-380.

——,* 1939. The Biology of Light Production in Arthropods. Smithsonian Inst. Year Book Ann. Rep. 1938, 377-404. Bibliography of 145 titles.

MILLER, G. 1935. Synchronous Firefly Flashing. Science, 81, 590-591.

MORRISON, T. F. 1929. Observations on the Synchronous Flashing of Fireflies in Siam. Science, 69, 400-401.

MORSE, E. S. 1924. The Synchronous Flashing of Fireflies. Science, 59, 163-164.

OKADA, YO K. 1935. Origin and Development of the Photogenic Organs of Lampyrids with Special Reference to Those of Luciola cruciata and Pyrocoelia rufa. Mem. Coll. Sc. Kyoto, 10B, 209-228.

——, 1928. Two Japanese Aquatic Glowworms. Trans. Entom. Soc. London, 74, 101-107.

PERKINS, M. 1931. Light of Glowworms. Nature, 128, 905.

RAMDAS, L. A. and L. P. VENKITESHWARAN. 1931. The Spectrum of a Glowworm (Lampyridae). Nature, 128, 726-727.

RAU, P. 1932. Rhythmic Periodicity and Synchronous Flashing in the Firefly, Photinus pyralis, with Notes on Photuris pennsylvanica. Ecology, 13, 7-11.

RICHMOND, C. A. 1930. Fireflies Flashing in Unison. Science, 71, 537-538.

RUEDEMANN, R. 1937. Observations on Excitation of Fireflies by Explosions. Science, 86, 222-223.

SMITH, H. M. 1935. Synchronous Flashing of Fireflies. Science, 82, 151-152.

SNELL, P. A. 1931, 1932.

SNYDER, C. D. and A. H. 1920. The Flashing Interval of Fireflies—Its Temperature Coefficient—An Explanation of Synchronous Flashing. Am. J. Physiol., 51, 536-543.

TAKAGI, S. 1934. Mitochondria in the Luminous Organs of Luciola cruciata, Motschulsky. Proc. Imp. Acad., 10, 692-694.

VERHOEFF, K. W. 1924. Zur Biologie der Lampyriden. Zeit. wiss. Insektenbiologie, 19.

VONWILLER, P. 1920. Anatomische Bemerkungen über den Bau der Leuchtorgane von Lampyris splendidula. Festschr. f. Zschokke No. 34. Basel.

Vogel, R. 1922. Über die Topographie der Leuchtorgane von Phausis splendidula, Leconte. Biol. Zentb., *42*, 138-140.

———, 1921. Bemerkungen zur Topographie und Anatomie der Leuchtorgane von Luciola chinensis, 1. Jen. Zeit. Naturw., *57*, 269-274.

Wood, R. W. 1939. A Firefly "Spinthariscope." Science, *90*, 233-234.

MOLLUSCA, BALANOGLOSSIDA AND TUNICATES

Berry, S. S. 1920. Light Production in Cephalopods. I and II. Biol. Bull., *38*, 141-195.

———, 1926. A Note on the Occurrence and Habits of a Luminous Squid (Abralia veranyi) at Madeira. Biol. Bull., *51*, 257-268.

Crozier, W. J. 1920. Persisting Rhythm of Light Production in Balanoglossids. Anat. Rec., *20*, 186.

Mortara, S. 1922. Sulla biofotogenesi. Rend. d R. Ac. Nat. d. Lincei, ser. 5, *31*, 54-58; 187-190.

Mortara, S. 1924. Sulla biofotogenesi e su alcuni batteri fotogeni. Riv. Biol., *6*, 1-22.

Naef, A. 1921. Die Zephalopoden. Fauna u. Flora der Golfes von Neapul. 35 Monog., 1st Part, Berlin. 148 pages.

Okada, Yo K. 1927. Luminescence chez les Mollusc Lamellibranches. Bull. Soc. Zool., France, *52*, 95-98.

———, 1927. Contribution a l'etude des Cephalopodes lumineus. Notes Preliminaires. Bull. Inst. Oceanog. Monaco No. *494*, 1-16; *499*, 1-15.

Okada, Yo K., S. Takagi and H. Sugino. 1933. Microchemical Studies on the So-called Photogenic Granules of Watasenia scintillans (Berry). Proc. Imp. Acad., *10*, 431-434.

Pierantoni, U. 1920. Gli organi luminosi simbiotici ed il loro circlo ereditario in Pyrosoma giganteum. Pub. Staz. Zool. Napoli, *3*.

———, 1920. Per una piu Osatta Conoscenza degli Organi Fotogeni dei Cefalopodi Abissali. Arch. Zool. Ital. *9*.

———, 1922. Simbiosi biofotogenesi. Att. d. R. ac. d. Lincei. Ren., *31*, 385.

———, 1924. Nuove Osservazioni su Luminescenzo e Simbiosi. III. Organo luminoso di Heteroteuthis dispar. Recon. d. R. Accad. Nat. d. Lincei, ser. 5, *33*, 61-65.

———, 1925. I Corpuscoli Fotogeni di Heteroteuthis dispar. Boll. Soc. Nat. Napoli, *38*.

———, 1926. Ancora sulla Bioluminescenza da Simbiosi. Riv. d. Biolog., *8*, 241-244.

Puntoni, V. 1925. La stato Atterale della Teoria Microbica della Biophotogenesi. Riv. di. Biol., *7*, 1-10.

Shima, G. 1927. Preliminary Note on the Nature of the Luminous Bodies of Watasenia scintillans (Berry). Proc. Imp. Acad. Tokyo, *3*, 461-464.

BIBLIOGRAPHY

SKOWRON, S. 1926. On the Luminescence of Some Cephalopods (Sepiola and Heteroteuthis). Riv. d. Biol., 8, 236-240.
STIER, A. 1938. Beitrage zur Embryonalentwicklung der Salpa pinnata. Zeit. f. Morph. u. Ökol der Tiere, 33, 582-632.
TAKAGI, S. 1933. Mitochondria in the Luminous Organs of Watasenia scintillans (Berry). Proc. Imp. Acad., 9, 651-654.
ZIRPOLO, G. 1922. Osservazioni sulla Biofotogenesi. Boll. Soc. Nat. Napoli, 34, 128-132.
———, 1923. Ricerche sulla Simbiosi fra Zoaxantelle e Phillirrhoë bucephala. Boll. Soc. Nat. Napoli, 35.

VERTEBRATA

AHL, E. 1929. Zur Kenntnis der Leuchtfische der Gatting, Myctophum. Zool. Anz., 81, 194-197.
BEEBE, WM. 1926. The Arcturus Adventure. Putnam. New York. pp. 215-219.
———, 1934. Half-mile Down. Harcourt Brace and Co. New York. See also, Nat. Geog. Mag., 59, 653-687, June, 1931; 61, 65-89, Jan. 1932 and 66, 661-704, Dec. 1934.
———, 1933-35. Deep-sea Fishes of the Bermuda Oceanographic Expeditions. Introduction. Family Alepocephalidea, Argentinidae, Idiacanthidae. Zoologica, 16, 5-241. Derichthyidae and Nessorhamphidae, 20, 1-51.
———, and J. CRANE. 1936-39. Deep-sea fishes of the Bermuda Oceanographic Expeditions. Family Serrivomeridae. Zoologica, 20, 53-102; Part II. Genus Platuronides, 22, 331-338, 1937; Family Nemichthyidae, 22, 349-383, 1937; Family Melanostomidae, 24, 66-238, 1939.
———, 1937. Preliminary List of Deep-sea Fish. Zoologica, 22, 197-208.
DAHLGREN, U. 1928. The Bacterial Light Organ of Ceratias. Science, 68, 65-66.
DEAN, B. Ed. by Gudger, E. W. and A. W. Heim. 1923. A Bibliography of Fishes. vol. 3, 513-514.
GIANFERRARI, L. 1922. Organi luminosi a bacteri nei pesci. Naturo, Padova, 13, 82-83.
GREENE, C. W. and H. H. 1924. Phosphorescence of Porichthys Notatus, the California Singing Fish. Am. J. Physiol., 70, 500-507.
HARMS, J. W. 1928. Bau und Entwicklung eines eigenartigen Leuchtorgans bei Equula spec. Zeit. wiss. Zool., 131, 157-179.
HARVEY, E. N. 1921, 1923, 1931.
HICKLING, C. F. 1925-26. A New Type of Luminescence in Fishes. J. Mar. Biol. Assn. (England), 13, 914-937; 14, 495-507.
———, 1928. The Luminescence of the Dog-fish, Spinax niger, Cloquet. Nature, 121, 280-281.
———, 1931. A New Type of Luminescence in Fishes. III. The Gland in Coelorhynchus coelorhynchus, Risso. J. Mar. Biol. Assn. U.K., 17, 853-875.

KISHITANI, T. 1930. Studien über die Leuchtsymbiose in Physiculus japonicus Hilgendorf, mit der Beilage der Zwei Neuen Arten der Leuchtbakterien. Sc. Rep. Tohoku Imp. Univ., 5, 801-823.

LADD-FRANKLIN, C. 1926. The Reddish Blue Arcs and the Reddish Blue Glow of the Retina; Seeing Your Own Nerve Currents Through Bioluminescence. Proc. Nat. Ac. Sc., 12, 413-414.

OKADA, YÔ K. 1926. On the Photogenic Organ of the Knight-fish (Monocentris Japonicus, Houttuyn). Biol. Bull., 50, 365-373.

PARKER, H. W. 1939. Luminous Organs in Lizards. J. Linn. Soc., 40, 658-660.

PIERANTONI, U. 1920. Organi Luminosi Batterici nei Pesci. Riv. di Biol., 3.

RAUTHER, M. 1927. Die Leuchtorgane. Bronns Klassen und Ordnung des Tierreiches, 6, abt. 1, book 2, 125-167. Leipzig.

SANZO, L. 1935. Uova, sviluppo embrionale stadi larvali post-larvali e giovanalidi di Sternoptychidae e Stomiotidae. 3. Maurolicus pennanti. R. Com. Talassogr. Ital. 2nd monog., 124-181.

SILVESTER, C. F. and FOWLER, H. W. 1926. A New Genus and Species of Phosphorescent fish, Kryptophanaron alfredi. Proc. Ac. Nat. Sc., Phila., 78, 245-247.

SKOWRON, S. 1928. Über das Leuchten der Tiefseefisches, Chauliodus Sloanii. Biol. Zentralb., 48, 680-684.

TROJAN, E. 1929. Die Geschlossenen Leuchtorgane der Tiefseefische. Congres. Intern. de Zool. a Budapest, Pt. 1, 743-747.

WATERMANN, T. H. 1939. Studies of Deep-sea Angler Fish. Bull. Mus. Comp. Zool., 85, 65-94.

YASAKI, Y. 1928. On the Nature of the Luminescence of the Knight fish (Monocentris japonicus, Houttuyn). J. Exp. Zool., 50, 495-505.

———, and Y. HANEDA. 1936. Ueber einen Neuen Typos von Leuchtorgan im Fische (Acropoma japonicum). Proc. Imp. Acad. (Tokyo), 12, 55-57.

BOOKS AND GENERAL WORKS ON LUMINESCENCE

* Papers marked with an asterisk contain bibliographies.

1936. First International Photoluminescence Congress in Warsaw. Acta Physica Polonica, 5, 1-431.

1939. Luminescence. Trans. Farad. Soc., 35, part 1, 240 pp.

BECKER,* J. A. 1923. Bibliography of Luminescence. Bull. Nat. Res. Coun., 5, 79-126. 1375 titles from 1906 to 1922, mostly non-biological.

BECQUEREL, E. 1867-1868. La Lumiere, ses Causes et ses Effects. Paris. T. 1.

BALY, E. C. C. 1915. A Theory of Absorption Fluorescence and Phosphorescence. Astrophys J., 42, 4-71.

CURIE, M. 1934. Luminescence des Corps Solides. Paris. 146 pp.

BIBLIOGRAPHY

DANKWORTT,* P. W.   1929. Lumineszenz-Analyse im Filtrierten Ultravioletten Licht. 2nd ed. Leipzig. 114 pp. 20 pl. 56 figs. Bibliography of 315 titles.

DESSAIGNE, J. PH.   1809. Memoire sur les Phosphorescences. J. de Physique, 68, 444-467, 1809; 69, 3-35, 169-200, 1809; 70, 109-128, 1810; 71, 67-70, 353-361, 1810; 73, 41-53, 1811; 74, 101-120, 173-193, 1812.

DHERE, C.   1933. Nachweis der Biologischeswichtigen Körper durch Fluoreszenzspektrens. Abderhaldens Handb. d. Biol. Arbeitsmethoden Abt. II. teil 3, Heft 4, lief. 420, pp., 3097-3306. 66 figs.

FONDA, G. R.   1938. The Fundamental Principles of Fluorescence. Trans. Amer. Inst. Elec. Engrs., 57, 5 pp.

GUDDEN, B.   1929. Phosphorescenz. In Müller-Pouillets Lehrbuch der Physik., 2, 2326-2349.

HAITINGER, M.   1938. Fluorescenz-Microskopie. Akad. Verlag. Leipzig. 108 pp.

———,   1937. Die Fluorescenz Analyse in der Mikrochemie. E. Hain. Wien. 192 pp.

HARVEY, E. N., E. Q. ADAMS, A. D. GARRISON, A. H. PFUND and H. S. TAYLOR.   1927. Chemiluminescence. Bull. Nat. Res. Council, No. 59, 7-62.

HEINRICH, J. P.   1811-1820. Die Phosphorescenz der Körper nach alle Umständen Untersucht und Erläutert. Nürnberg bei Schrag. 596 pp.
1st Abhandlung 1811 Phosphorescenz durch Licht, 1-132.
2nd Abhandlung 1812 Phosphorescenz durch Temperaturerhohlung, 133-312.
3rd Abhandlung 1815 Das Leuchten des Pflanzen und Thieren, 313-424.
4th Abhandlung 1820 Phosphorescenz durch Mechanische Erregung, 425-570.
5th Abhandlung 1820 Leuchten bei Chemischer Mischerung, 573-596. Also Schweigger's J. f. Chem. u. Physik., 30, 218-239, 1820.

KAUFFMANN, H.   1925. Methoden zur Untersuchung von Fluorescenzerscheinungen. Abderhalden's Handb. d. biol. Arbeitsm. Ab. 2, teil 1, 131-170.

KAYSER,* H.   1908. Handbuch der Spectroskopie. Leipzig. Phosphorescence, 4, Chap. 5, 599-838. Fluorescence, 4, Chap. 6, 841-1214. References to all the older literature.

LENARD,* P., F. SCHMIDT and R. TOMASCHEK.   1928. Phosphorescence and Fluorescenz. Handb. d. Exper. Physik of W. Wien and F. Harms., 23, 1-1041. Many references to literature.

MERRITT, E., E. L. NICHOLS and C. D. CHILD.   1923. Selected Topics in the Field of Luminescence. Bull. Nat. Res. Coun. No. 5, 1-126.

NICHOLS, E. L., H. L. HOWES and D. T. WILBER.   1928. Cathodoluminescence and the Luminescence of Incandescent Solids. Pub. Carn. Inst. Wash. No. 384. 350 pp. 176 figs.

NICHOLS, E. L. and E. MERRITT. 1912. Studies in Luminescence. Pub. Carn. Inst. Wash. No. 152, 1-223.
POHL, R. W. 1929. Fluorescenz. In Müller-Pouillet's Lehrbuch der Physik, 2, 2311-2325.
PRIESTLEY, J. 1772. The History and Present State of Discoveries Relating to Vision, Light and Colours. London. 1, 360-383.
PRINGSHEIM,* P. 1928. Fluorescenz und Phosphorescenz. Julius Springer. Berlin. 3rd ed. 357 pp. Bibliography of 637 titles, 1908 to 1927.
RADLEY,* J. A. and J. GRANT. 1933. Fluorescence Analysis in Ultraviolet Light. 219 pp. 15 figs. London. Chapman and Hall. 1597 references (at end of chapters).
RUPP, H. 1937. Die Leuchtmassen und Ihre Verwendung. Berlin.
TOMASCHEK, R. 1928. Phosphorescenz, Fluorescenz und Chemisches Reactionsleuchten. In E. Gehrcke's Handb. d. Physikalischen Optik, 2, 229-358.
VANINO, L. 1935. Die Leuchtfarbe. Stuttgart.
WEISER, H. B. 1928. The Phosphorescent Sulphides as Colloid Systems. In the Colloid Salts (McGraw Hill) Chap. 10, 156-174.
WOOD, R. 1934. Physical Optics. MacMillan and Co. 3rd ed. Chap. 18 and 19. Resonance Radiations and Fluorescence, 587-647. Chap. 20. Fluorescence and Phosphorescence of Solids and Liquids, 648-667.

THERMOLUMINESCENCE, PHOSPHORESCENCE AND FLUORESCENCE

ANDREWS, W. S. 1920. A Special Form of Phosphoroscope. Gen. Elec. Rev., 23, 855-857.
ARMSTRONG, H. E. 1924. Problems of Hydrone, etc. Luminous Ice. Nature, 113, 163.
BERNARD, J. E. and F. V. WELCH. 1936. Fluorescence Microscopy With High Powers. J. Roy. Mic. Soc., 56, 361-371.
COCKAYNE, E. A. 1924. The Distribution of Fluorescent Pigments in Lepidoptera. Trans. Ent. Soc. London. 19 pp. Aug. 1924.
COOLIDGE, W. D. 1925. Modern X-ray Tube Development. J. Frank. Inst., 201, 619-648.
COOLIDGE, W. D. and C. N. MOORE. 1926. The Production of Highvoltage Cathode Rays Outside of the Generating Tube. J. Frank. Inst., 202, 693-734. Also Gen. Elec. Rev., 35, 413-417, 1932.
FONDA, G. R. 1939. Characteristics of Silicate Phosphores. J. Phys. Chem., 43, 561-577. Also, J. App. Physics, 10, 408-420.
GIESE, A. C. and P. A. LEIGHTON. 1933. Fluorescence of Cells in the Ultra Violet. Science, 77, 507-510; on phosphorescence, 85, 428-429, 1937.
HAMPERL, H. 1934. Die Fluorescenzmikroskopie Menschlicher Gewebe. Arch. Path. Anat. u. Physiol., 292, 1-51.
HARVEY, E. N. 1926.

BIBLIOGRAPHY

JOHNSON, R. P. 1939. Luminescence of Sulphide and Silicate Phosphors. J. Opt. Soc. Amer., 29, 283-289, 387-396.

LLOYD, F. E. 1924. The Fluorescent Colors of Plants. Science, 59, 241-248; also 58, 91-92; 229-230, 1923.

NEWCOMER, H. S. 1920. The X-ray Fluorescence of Certain Organic Compounds. J. Am. Chem. Soc., 42, 1997-2007.

SEITZ, F. 1938. Interpretation of the Properties of Alkali Halide-Thallium Phosphors. J. Chem. Phys., 6, 150-162; also, Zinc Sulphide Phosphors, 454-461.

SINGER, E. 1932. A Microscope for Observation of Fluorescence in Tissues. Science, 75, 289-291.

———, 1933. Observations on the Frog's Kidney with the Fluorescence Microscope. Am. J. Anat., 53, 469-488; on leucocytes, Anat. Rec., 58, 93-99, 1933.

WICK, F. G. 1927. The Effect of X-rays on Thermoluminescence. J. Opt. Soc. Amer. and Rev. Sci. Inst., 14, 33-44; 125-132; 16, 398-408. See also 21, 223-231, 1931.

ELECTROLUMINESCENCE AND SONOLUMINESCENCE

CHAMBERS, L. A. 1937. The Emission of Visible Light From Cavitated Liquids. J. Chem. Phys., 5, 290-292.

DUSHMAN, S. 1934. Production of Light from Discharges in Gasses. Gen. Elec. Rev., 37, 260-268.

FLOSDORF, E. W., L. A. CHAMBERS and W. M. MALISOFF. 1936. Sonic Activation in Chemical Systems; Oxidations at Audible Frequencies. J. Am. Chem. Soc., 58, 1069-.

FRENZEL, J. and H. SCHULTES. 1934. Luminescenz in Ultra-schallbeschickten Wasser. Zeit. physik Chemie, 27, 421-424.

HARVEY, E. N. 1939.

LEVSIN, V. L. and S. N. KZEVKIN. 1937. On the Mechanism of Luminescence in Liquids under Ultrasonic Treatment. C.R. Ac. Sc. U.S.S.R., 16, 399-404.

POLOTSKII, I. G. 1938. J. Gen. Chem. (U.S.S.R.) 8, 1691-5.

TRIBOLUMINESCENCE AND CRYSTALLOLUMINESCENCE

HARVEY, E. N. 1939.

INONE, T., Kunitomi and E. Shibata. 1939. Über Triboluminescenz. J. Sci. Hiroshima U. series A 9, 129-137.

NELSON, D. M. 1926. Photographic Spectra of Triboluminescence. J. Opt. Soc. Am., 12, 207-215.

TUDA, S., T. TAKETA and E. SHIBATA. 1939. Über Kristalloluminescenz. J. Sci. Hiroshima Univ., series A, 9, 137-144.

WICK, F. G. 1937. An Experimental Study of the Triboluminescence of Certain Natural Crystals and Synthetically Prepared Materials. J. Opt. Soc. Am., 27, 275-285.

WICK, F. G. 1939. Effect of Temperature and Exposure to X-rays upon Triboluminescence. J. Opt. Soc. Am., 10, 407-412.

CHEMILUMINESCENCE AND GALVANOLUMINESCENCE

ADAMS, E. Q. 1924. The Luminous Efficiency of Chemiluminescent Reactions. Phys. Rev., 23, 771.

———, 1927. Energetics, Spectra, Intensity and Efficiency of Chemiluminescence. Bull. Nat. Res. Coun. No. 59, 30-40.

ALBRECHT, H. O. 1928. Über die Chemiluminescenz des Aminophthalsäurehydrazids. Z. f. Phys. Chem., 136, 321-330.

BOEHMANN, W. E. 1934. Synthesis of Phenanthrene Derivatives. I. Reactions of 9 Phenanthrylmagnesium Bromide. J. Am. Chem. Soc., 56, 1363-1367.

BAWN, C. E. H. and W. J. DUNNING. 1938. Chemiluminescence of Sodium Vapor with Organic Halides. Trans. Farad. Soc., 35, 185-190.

BHATNAGAR, S. S. and K. C. MATHUR. 1932. Die Chemiluminescence von Amarin. Zeit. Physik. Chem. A 159, 454-458.

DECKER, H. and W. PETSCH. 1935. Biacridyl und die sich von ihm Ableitenden Radikale und Leuchtsalze, die Luzigenine. J. f. Prak. Chemie, 143, 7-9, 211-235.

DOWNEY, W. E. 1924. The Relation Between the Glow of Phosphorus and the Formation of Ozone. J. Chem. Soc., 125, 347-357.

DREW, H. D. K. and collaborators. 1937. Chemiluminescent Organic Compounds (Phthalhydrazids). J. Chem. Soc., 1937, 16-37; 586-592; 1841-1846; 1938, 791-793. Also, Trans. Farad. Soc., 35, 207-215, 1938.

DUFFORD, R. T. 1929. Luminescence Associated with Electrolysis. J. Opt. Soc. Am. and Rev. Sci. Inst., 18, 17-28.

DUFFORD, R. T., S. CALVERT and D. NIGHTINGALE. 1923. Luminescence of Organo-Magnesium Halides. J. Am. Chem. Soc., 45, 278-285; 2058-2072.

DUFFORD, R. T., D. NIGHTINGALE and S. CALVERT. 1924. Spectra of Luminescence of Grignard Compounds. J. Opt. Soc. Am. and R.S.I., 9, 405.

———, 1925. Luminescence of Grignard Compounds: Spectra and Brightness. J. Am. Chem. Soc., 47, 95-102.

DUFFORD, R. T., D. NIGHTINGALE and L. W. GADDUM. 1937. Luminescence of Grignard Compounds in Electric and Magnetic Fields and Related Electrical Phenomena. J. Am. Chem. Soc., 49, 1858-1864.

EVANS, M. G. and M. POLANYI. 1938. Notes on the Luminescence of Sodium Vapor in Highly Dilute Flames. Trans. Farad. Soc., 35, 178-184.

EVANS, M. G., H. EYRING and J. F. KINCAID. 1938. Non-adiabatic Reactions. Chemiluminescence. J. Chem. Physics, 6, 349-358.

BIBLIOGRAPHY

Evans, W. V. and R. I. Dufford. 1923. Luminescence of Compounds Formed by the Action of Magnesium on Paradibromo-benzene and Related Compounds. J. Am. Chem. Soc., 45, 278-285.

Evans, W. V. and E. M. Diepenhorst. 1926. Further Studies on Luminescent Grignard Compounds. J. Am. Chem. Soc., 48, 715-724.

Garrison, A. D. 1927. Types of Chemiluminescence. Bull. Nat. Res. Coun. No. 59, 7-22.

Gleu, K. and W. Petsch. 1935. Die Chemiluminescenz der Dimethyldiacridyliumsalze. Angewandte Chemie, 48, 57-59.

Gleu, K. and K. Pfannstiel. 1936. Über 3-aminophthalsäure-hydrazid. J. f. Prakt. Chem., 146, 137-150.

Haber, F. and W. Zisch. 1922. Anregung von Gasspektren durch Chemischen Reaktionen. Zeit. f. Physik, 9, 302-326.

Harris, L. and A. S. Parker. 1939. The Chemiluminescence of 3-Aminophthalhydrazide. J. Am. Chem. Soc., 57, 1939-1942.

Harvey, E. N. 1916, 1917, 1927, 1929.

Helberger, J. H. 1938. Über eine neuen Fall von Chemiluminescence. Naturwisschen., 26, 316-317.

Huntress, E. H., L. N. Stanley and A. S. Parker. 1934. The Preparation of 3-Aminophthalhydrazide for Use in the Demonstration of Chemiluminescence. J. Am. Chem. Soc., 56, 241-242.

Kautsky, H. 1921. Über einige Ungesättigte Siliciumverbindungen. Zeit. f. Amorg. u. Alleg. Chem., 117, 209-242.

———, 1938. Quenching of Luminescence by Oxygen. Trans. Farad. Soc., 35, 216-218.

Kautsky, H. and H. Hohn. 1936. Energieumwandlungen an Grenzflächen. Lumineszenznachweis Tautomerer Formen Adsorbierter Moleküle. Kolloidzeit., 75, 164-169.

Kautsky, H. and O. Neitzke. 1925. Spektren Emissionsfähiger Stoffe bei Erregung durch Licht und durch Chemische Reactionen. Zeit. f. Physik., 31, 60-71.

Kautsky, H. and H. Thiele. 1925. Umsetzungen des Siloxens mit Halogenverbindungen und ihre Auslösung durch Licht und Chemische Reaktionen. Zeit. anorg. u. Alleg. Chem., 144, 197-217.

Kautsky, H. u. H. Zocher. 1922. Über die Beziehung zwischen Chemi- und Photolumineszenz bei Unsättigten Siliziumverbindungen. Zeit. f. Physik, 9, 267-284.

Lifschitz, J. and O. E. Kalberer. 1922. Über Chemilumineszenz und Thermochemisches Verhalten von Organomagnesiumverbindungen. Zeit. Physik Chem., 102, 393-415.

Petrikaln, A. 1924. Über die Chemilumineszenz und die Energieumwandlung bei der Oxydation des Phosphors. Zeit. Physik 22, 119-126.

POLANYI, M. and collaborators. 1928. Über Hochverdünnte Flammen. Zeit. Physik Chem. B. 1, 3-73.
RAYLEIGH, LORD. 1921. The Glow of Phosphorus. Proc. Roy. Soc., 99, 372-384. See Nature, 114, 612-614, 1924.
SCHALES, O. 1939. Die Katalytisches Beeinflussung der Luminescenz des 3-Aminophthalsäurehydrazids durch Häemine und Häeminderivate. Ber. d.d. Chem. Ges., 72, 167-177; also, 71, 447, 1938.
STUCHTHEY, K. 1920. Spectralanalytische Untersuchungen des Leuchtens Zerfallenden Ozons. Zeit. f. wiss. Phot., 19, 161-197.
SPONER, H. 1936. Molekülspektren. Chemiluminescence, pp. 450-461. II. Text. 506 pp. Berlin. J. Springer.
SVESNIKOV, B. J. 1938. On the Mechanism of the Chemiluminescence of 3-Amino-phthalic Hydrazide. Acta Physicochimica, U.S.S.R., 8, 441-460.
TAYLOR, H. S. 1927. Photochemistry and Chemiluminescence. Kinetics of Chemiluminescence. Bull. Nat. Res. Coun. No. 59, 41-49.
TAMAMUSHI, B. 1937. Die Anregung des Leuchten von 3-Aminophthalsäurehydrazid durch Molekülaren Säuerstoff und Haemin as Katalysator. Naturwiss., 25, 318.
TAMAMUSHI, B. and H. AKIYAMA. 1938. Zum Mechanismus der Chemiluminescenz des 3-Aminophthalsäurehydrazid. Z. Physik. Chem. B 38, 400-406.
THIELERT, H. and P. PFEIFFER. 1928. Zur Kenntniss der Lumineszenzercheinungen bei der Oxydation des Luminols. B. d. d. Chem. Ges., 71, 1399-1403.
THOMAS, C. D. and R. T. DUFFORD. 1933. Efficiency of the Chemiluminescence Accompanying Oxidation of Grignard Compounds. J. Op. Soc. Amer., 23, 251-255.
TRAUTZ, M. and W. SEIDELL. 1922. Über die Lumineszenz zerfallenden Ozons. Ann. Physik., 67, 527-572.
WEISER, H. B. and A. GARRISON. 1921. The Oxidation and Luminescence of Phosphorus. I, II, III. J. Physik. Chem., 25, 61-81; 349-384; 473-490.
WEISS, J. 1938. Oxidation and Chemiluminescence. Trans. Farad. Soc., 35, 207-214.
WITTE, A. A. M. 1935. Luminescence Phenomena with Benz-hydrazides and Benzenesulphon-hydrazides. Rev. trav. Chim., 54, 471-475.
ZELLNER, C. N. and G. DOUGHERTY. 1937. The Chemiluminescence of Phthalhydrazide Derivatives. J. Am. Chem. Soc., 59, 2580-2583.
ZOCHER, H. and H. KAUTSKY. 1923. Über Lumineszenz bei Chemische Reaktionen. Die Naturwissensch. 11, 194-199; also Zeit. f. Electroc., 29, 308, 1923.

BIBLIOGRAPHY

INVISIBLE AND PENETRATING RADIATION; MITOGENETIC RAYS

AUDUBERT, R.   1936. Emission de Rayonnement par les Reactions Chimiques. J. Chem. Phys., 33, 507-525; also Trans. Farad. Soc., 35, 197-206, 1938.
BATEMAN, J. B.   1935. Mitogenetic Radiation. Biol. Rev., 10, 42-71.
GURWITSCH, A.   1923. Die Nature des Specifischen Erregers der Zellteilung. Arch. Entw., 100, 11-40; 103, 490-498.
―――, 1932. Die Mitogenetische Strahlung. Berlin. J. Springer. 1932.
GURWITSCH, A. and L. GURWITSCH.   1925. Über der Ursprung der Mitogenetischen Strahlen. Arch. Entw., 105, 470-472; 473-474.
―――, 1937. Mitogenetic Analysis of the Excitation of the Nervous System. Amsterdam. 100 pp. 30 figs.
HOLLAENDER, A.   1936. The Problem of Mitogenetic Rays; in Biological Effects of Radiation. ed. by B. M. Duggar, New York, 2, 919-960.
HOLLAENDER, A. and W. D. CLAUS.   1937. An Experimental Study of the Problem of Mitogenetic Radiation. Bull. Nat. Res. Coun., No. 100, 96 pp.
HOLLAENDER, A. and E. SCHOEFFEL.   1931. Mitogenetic Rays. Qu. Rev. Biol., 6, 215-222.
KEENAN, G. L.   1926. Substances Which Affect Photographic Plates in the Dark. Chem. Rev., 3, 95-111.
KUGELMASS, I. N. and I. MCQUARRIE.   1925. "Russel Effect" not Ultraviolet Light responsible for Changes produced in the Photographic Plate by Antirachitic Substances. Science, 62, 87-88; also 60, 272-273.
RAHN, O.   1936. Invisible Radiations of Organisms. Berlin. Geb. Borntrager. 215 pp.
REITER, O. and D. GABOR.   1928. Zellteilung und Strahlung. Sonderheft d. Wiss. Veroeff. aus dem Siemens-Konzern. J. Springer. Berlin.
TAYLOR, G. W. and E. N. HARVEY.   1931. The Theory of Mitogenetic Radiation. Biol. Bull., 61, 280-293.

LIGHT MEASUREMENT AND EFFICIENCY

BRACKETT, F. S.   1937. Measurement and Application of Visible and Near Visible Radiation; in Biological Effects of Radiation. ed. by B. M. Duggar. Paper IV, 1, 123-210. McGraw Hill Book Co., New York.
CADY, F. C. and H. B. DATES, ed.   1925. Illuminating Engineering. John Wiley and Sons. New York. 486 pp.
DUSHMAN, S. 1937. The Search for High Efficiency Light Sources. J. Opt. Soc. Amer., 27, 1-24.
―――, 1938. Recent Developments in Gaseous Discharge Lamps. J. Soc. Motion Pic. Engin., 30, 58-80.

FORSYTHE, W. E. 1935. The Present Status of Photometry. Trans. Ill. Eng. Soc. 16 pp.

———, (editor). 1937. Measurement of Radiant Energy. McGraw Hill Book Co., New York. 21 contributors. 452 pp.

FOUND, C. G. 1937. Fundamentals of Electric Discharge Lamps. Trans. Ill. Eng. Soc., 29 pp.

MOON, P. 1936. The Scientific Basis of Illuminating Engineering. New York. McGraw Hill Book Co. 597 pp.

PFUND, A. H. 1927. Experimental Methods Used in the Study of Luminescence. Bull. Nat. Res. Coun., No. 59, 23-29.

WALSH, J. W. T. 1926. Photometry. New York. Van Nostrand Co. 505 pp.

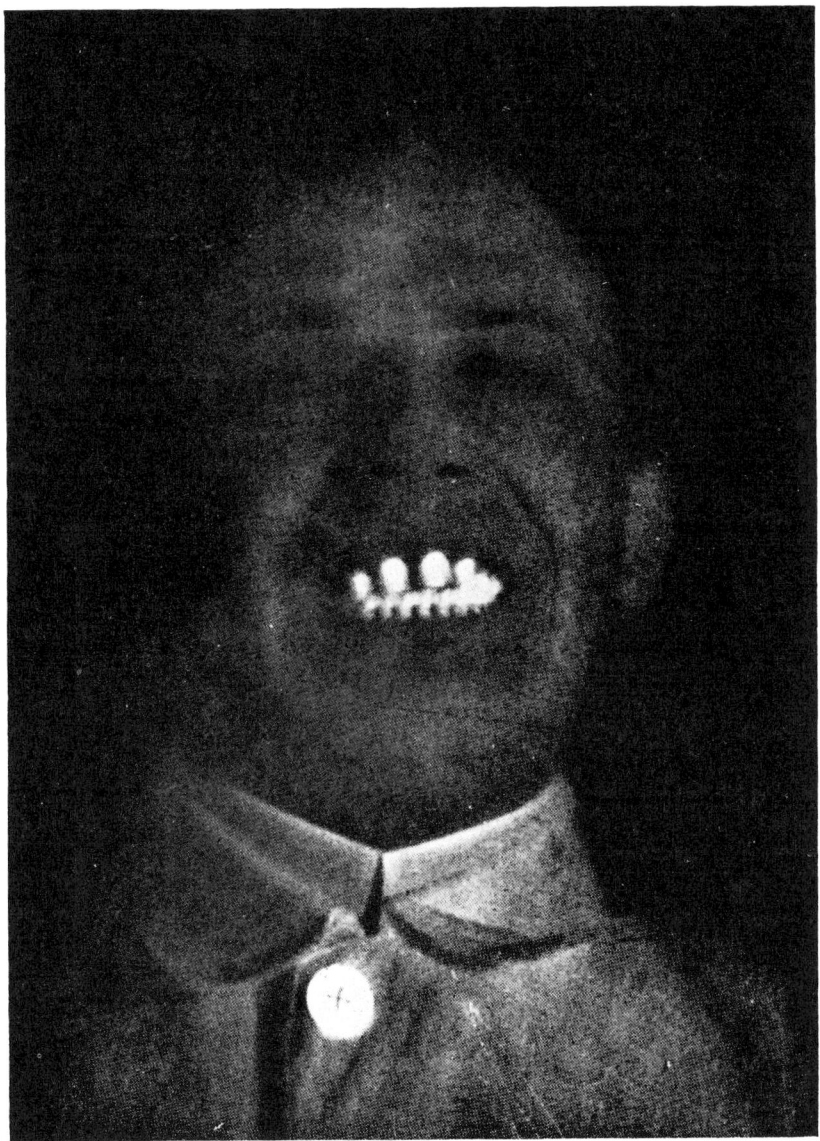

Fig. 1 Man's face, photographed by fluorescent light excited by ultraviolet without visible rays. Note that teeth and bone button are especially bright. From Dankwortt's *Luminescenz-analyse*, by courtesy of Akademische Gesellschaft, Leipzig.

FIG. 2 Two dead herring photographed by the light of luminous bacteria growing on them. (After Pratje, from a photograph of Professor Rosen.)

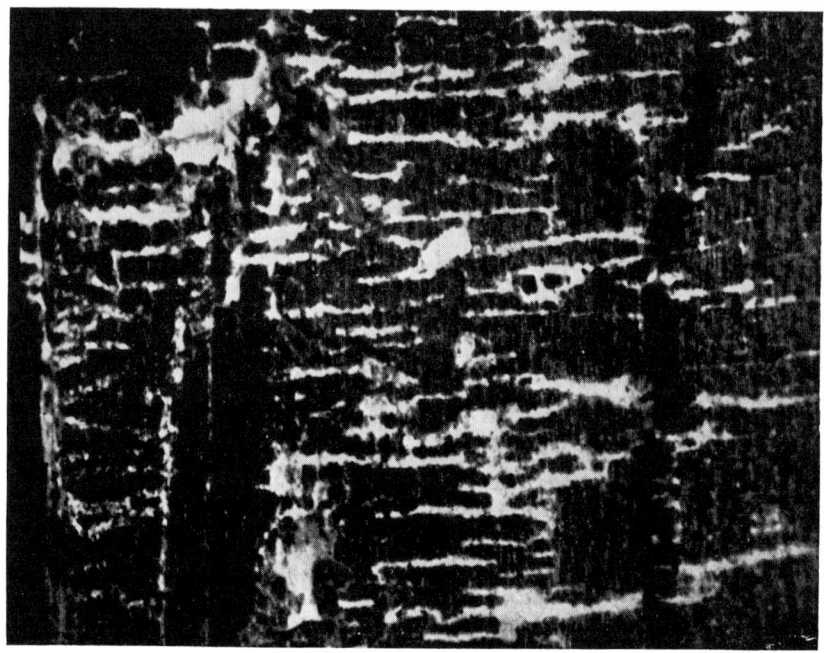

FIG. 3 Spruce slab infiltrated with the mycelium of a fungus, often responsible for "shining wood." From Atkinson's *Mushrooms* by courtesy of Henry Holt and Co.

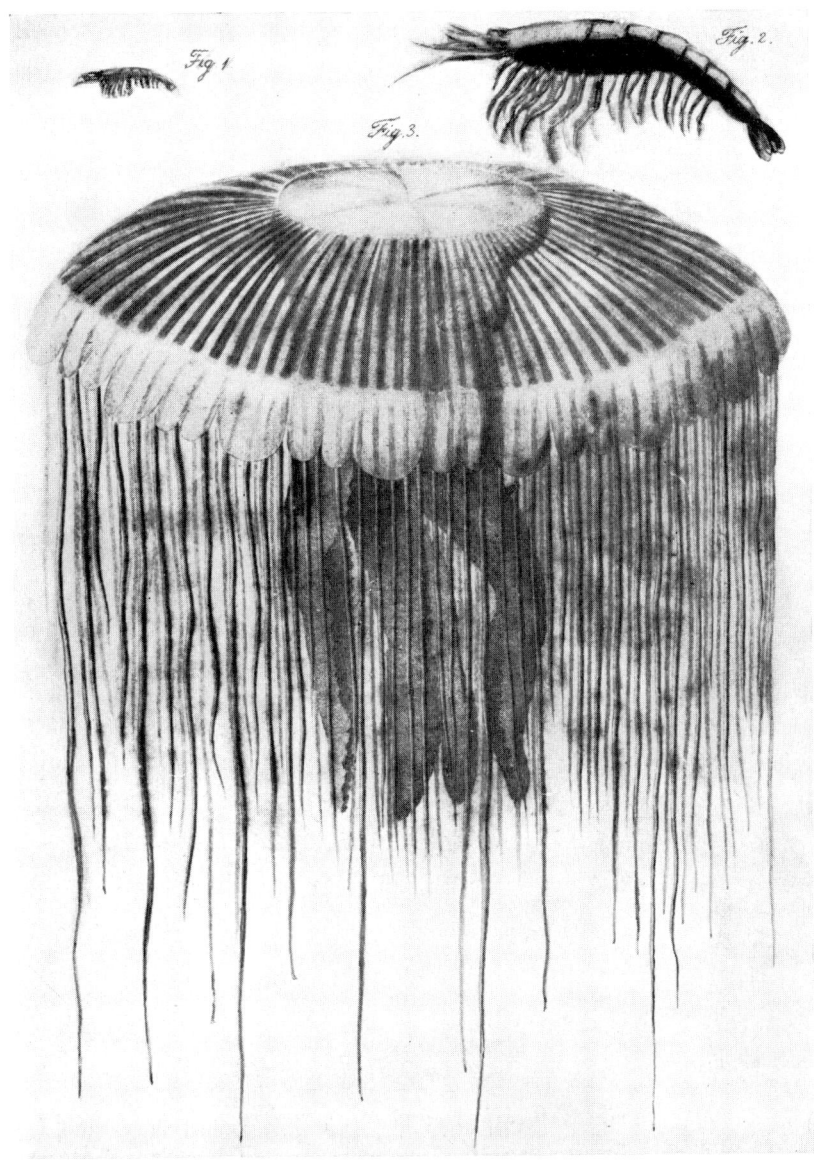

FIG. 4 Large luminous animals of the sea, two shrimp and a jellyfish. From an old print by Macartney, 1810.

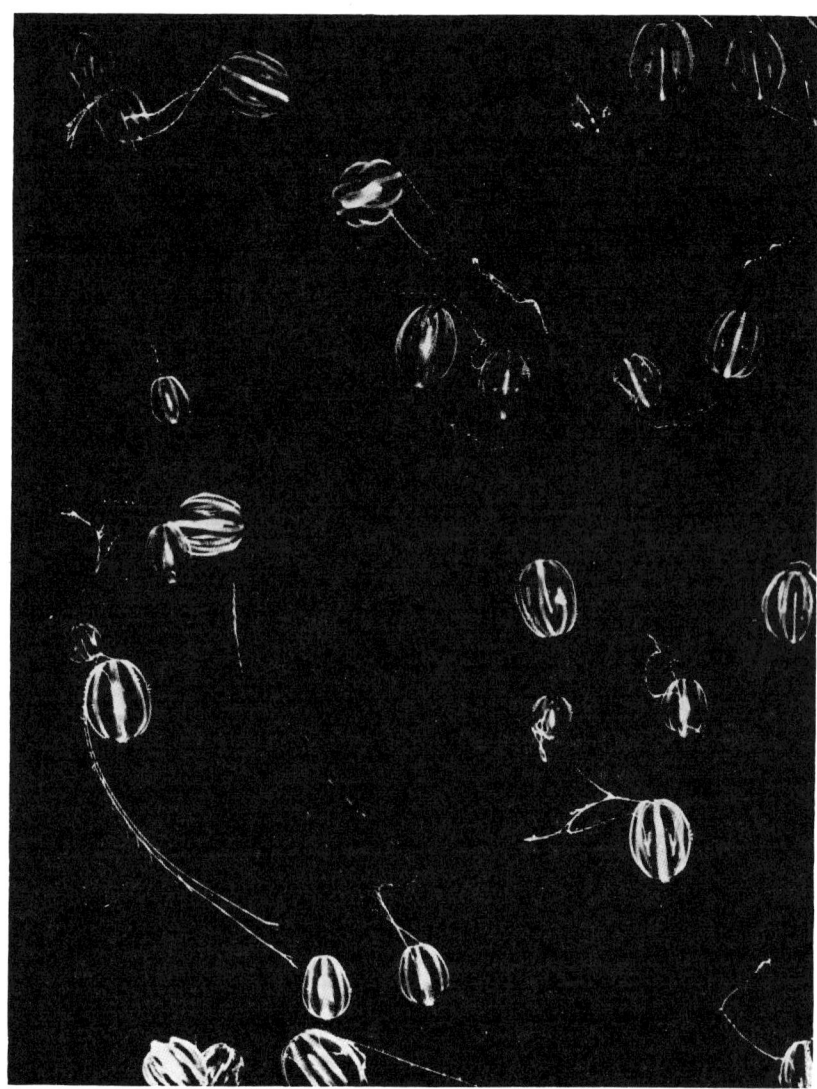

Fig. 5 Comb-jellies (*Pleurobrachia*), responsible for much of the larger flashes of light observed in the sea at night. From Buchsbaum's *Animals without Backbones*, by courtesy of Chicago University Press from a photograph by F. Schensky.

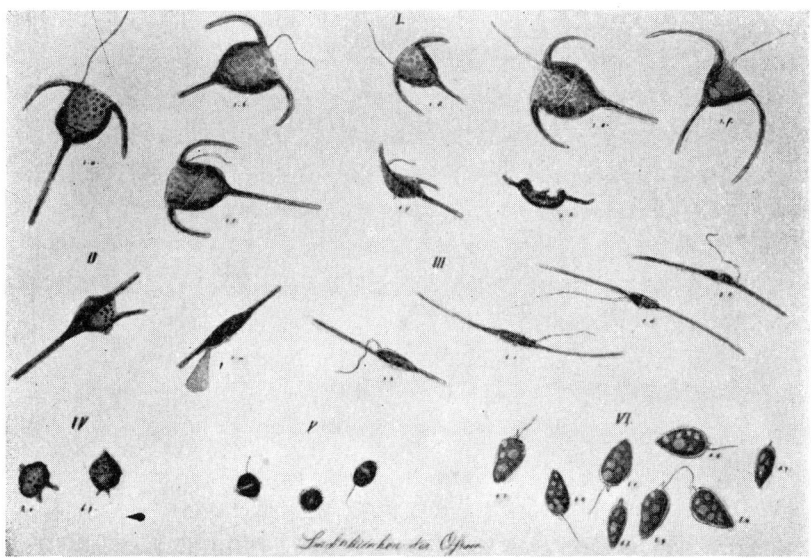

Fig. 6 Microscopic organisms responsible for "burning of the sea." Above. Dinoflagellates, after Ehrenberg, 1834. Below. *Noctiluca*, after Quatrefages, 1850, both highly magnified. Note the points of luminescence due to granules within *Noctiluca* and (at right) the light from a fragment.

FIG. 7 A fish of the Banda Sea, Photoblepharon palpebratus, whose light organs under each eye harbor symbiotic luminous bacteria, the real source of the light. Reproduced by kindness of Ulric Dahlgren, after a drawing by Bruce Horsfall.

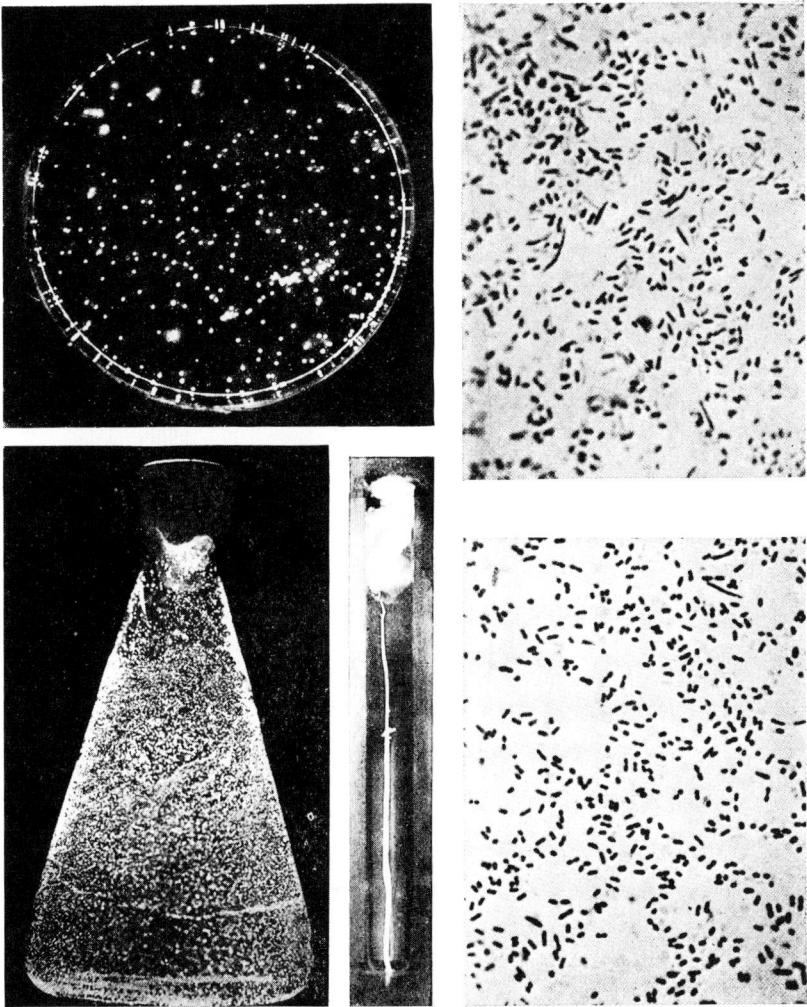

Fig. 8 Left. Luminous bacterial colonies growing on culture media in a Petri dish, flask and test tube (after Molish). Right. Individual bacteria stained on a slide and highly magnified; above, *Pseudomonas Toyamensis*; below, *Coccobacillus Ikiensis*, after Kishitani.

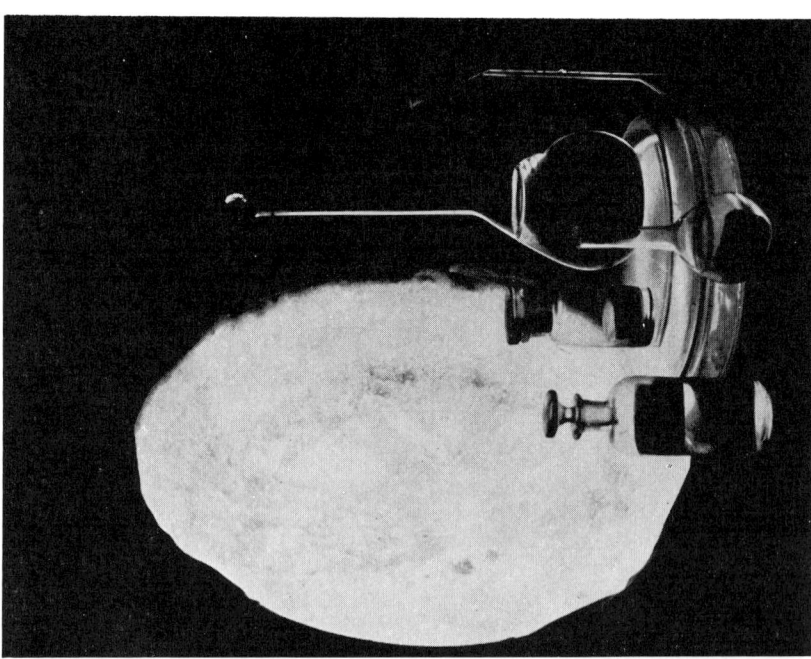
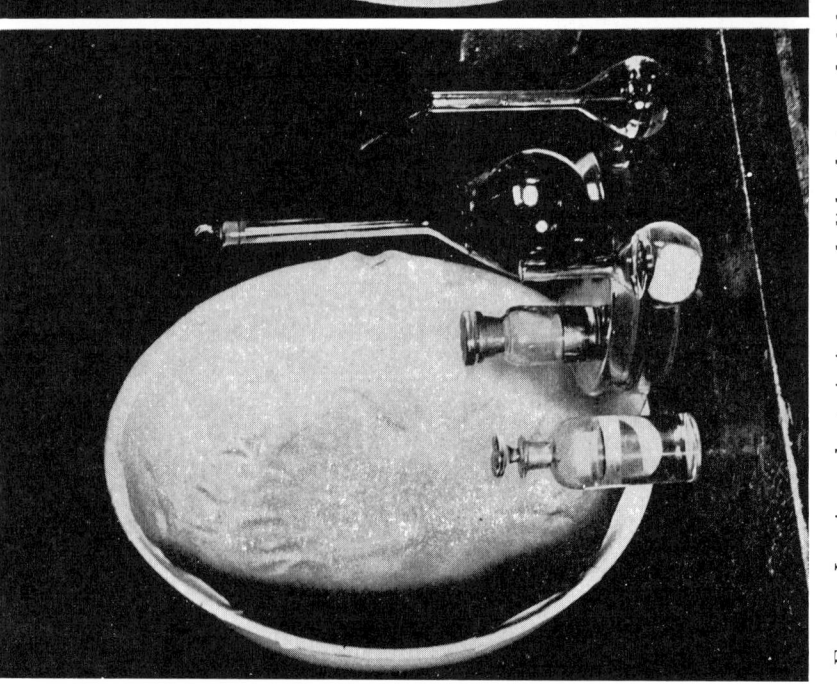

Fig. 9. Luminous bacteria in a round dish photographed by their own light (right) and by ordinary light (left). Note that labels on bottles cast a shadow with light from bacteria and the difference in high lights on glassware in each picture. Bacteria prepared by Dr. F. H. Johnson, photographed by Goro, Black Star, reproduced by courtesy of F. Goro and *Life*.

FIG. 10 A sand-flea such as is often afflicted by a luminous bacterial malady, that is always fatal. From Buchsbaum's *Animals without Backbones*, by courtesy of Chicago University Press and American Nature Association.

FIG. 11 The mycelium of *Armillaria mellea*, a luminous fungus found growing on wooden supports of coal mines. From Atkinson's *Mushrooms*, by courtesy of Henry Holt and Co.

Fig. 12 Luminous fishes of the Banda Islands. Two *Photoblepharon* (above) and two *Anomalops* (below), from an original photograph. Note that one in each pair has its light organ covered.

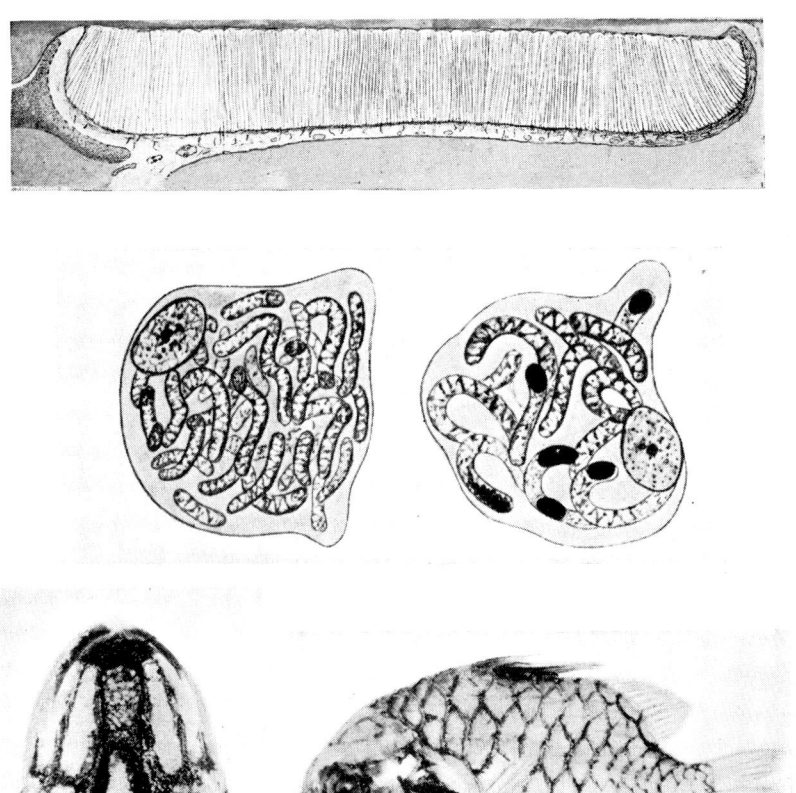

Fig. 13 (above) A cross section of the luminous organ of *Photoblepharon*, showing the tubes containing luminous bacteria. A rich supply of blood capillaries run between the tubes. After Steche.

Fig. 14 (center) Structures in the luminous cells of *Pyrosoma*, believed to be symbiotic bacteria. Note the black bodies, possibly spores. After Pierantoni.

Fig. 15 (below) The fish, *Monocentris japonica*, whose lower jaw contains a luminous organ in which grow luminous bacteria. The figure at left is a ventral view of the jaw. The black spot at tip is luminous. After Yasaki.

Fig. 16 A Japanese fish, *Physiculus japonicus*, with a large internal gland containing luminous bacteria which can be ejected through the opening, S. A, anus. O, urogenital opening. After Kishitani.

FIG. 17 A mushroom, *Cliocybe illudens*, in which the fruiting body is luminous. After Atkinson's *Mushrooms*, by courtesy of Henry Holt and Co.

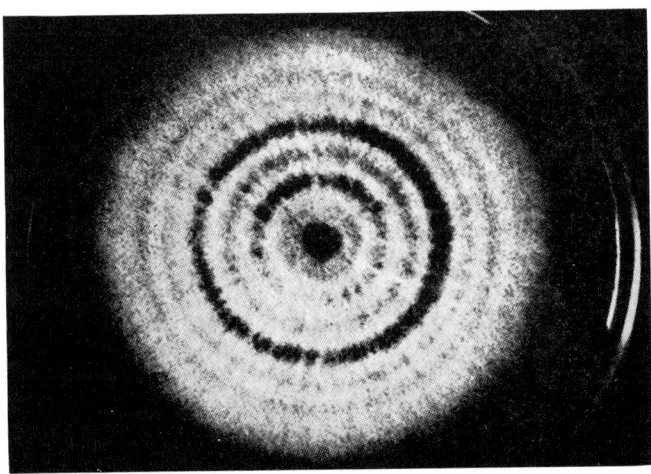

FIG. 18 The luminous mycelium of the coffee leaf spot fungus, *Omphalia flavida*, growing in a Petri dish. The concentric rings are due to formation of black gemmifer bearing mycelium during the day, white mycelium at night. After Buller's *Researches on Fungi*, by courtesy of Longmans Green and Co. and A. H. R. Buller.

Fig. 19 The luminous leaf spot fungus, *Omphalia flavida*, on a coffee leaf. This particular fungus can be easily identified on coffee trees on dark nights by its luminescence.

Fig. 20 The stag-horn fungus, *Xylaria hypoxylon*, reported by some to be luminous. Figs. 19 and 20 from Buller's *Researches on Fungi*, by courtesy of Longmans Green and Co. and A. H. R. Buller.

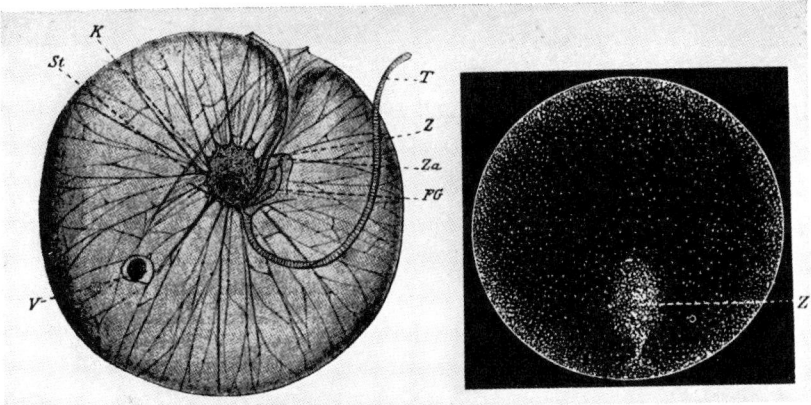

Fig. 21 A modern figure of Noctiluca, by day (left) and at night (right). Note tentacle and central region, Z, where luminous granules are most concentrated. After Pratje.

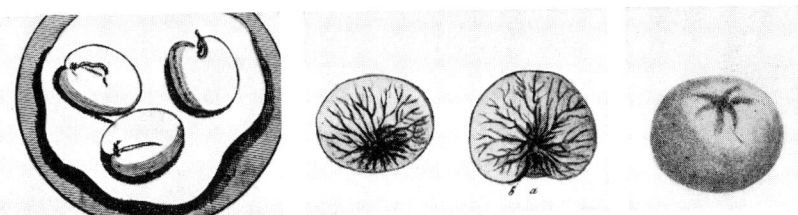

Fig. 22 The oldest figures of Noctiluca; three in dish at left, after Dicquemare, 1775; two in center after Slabber, 1778; one at right after Macartney, 1810.

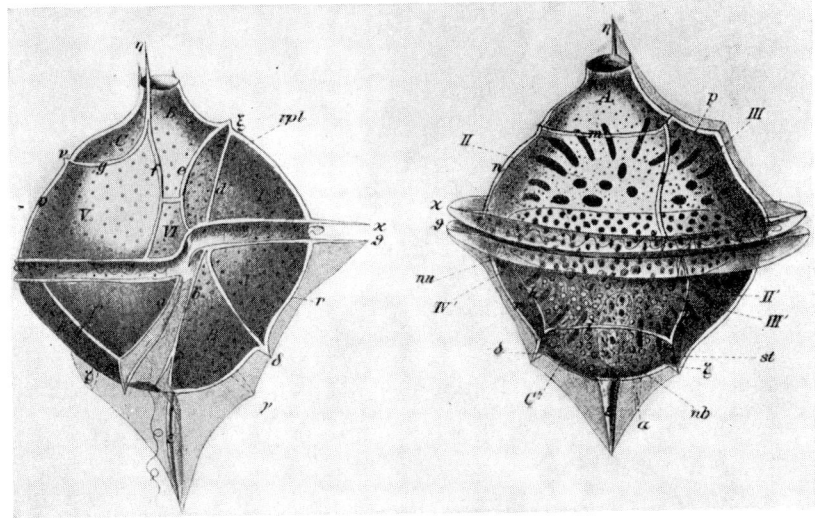

Fig. 23 Dinoflagellates (Peridinium bahamense), responsible for the luminescence of "Fire Lake" in the Bahamas. Note the plates, which make a protective covering for these forms. After Plate.

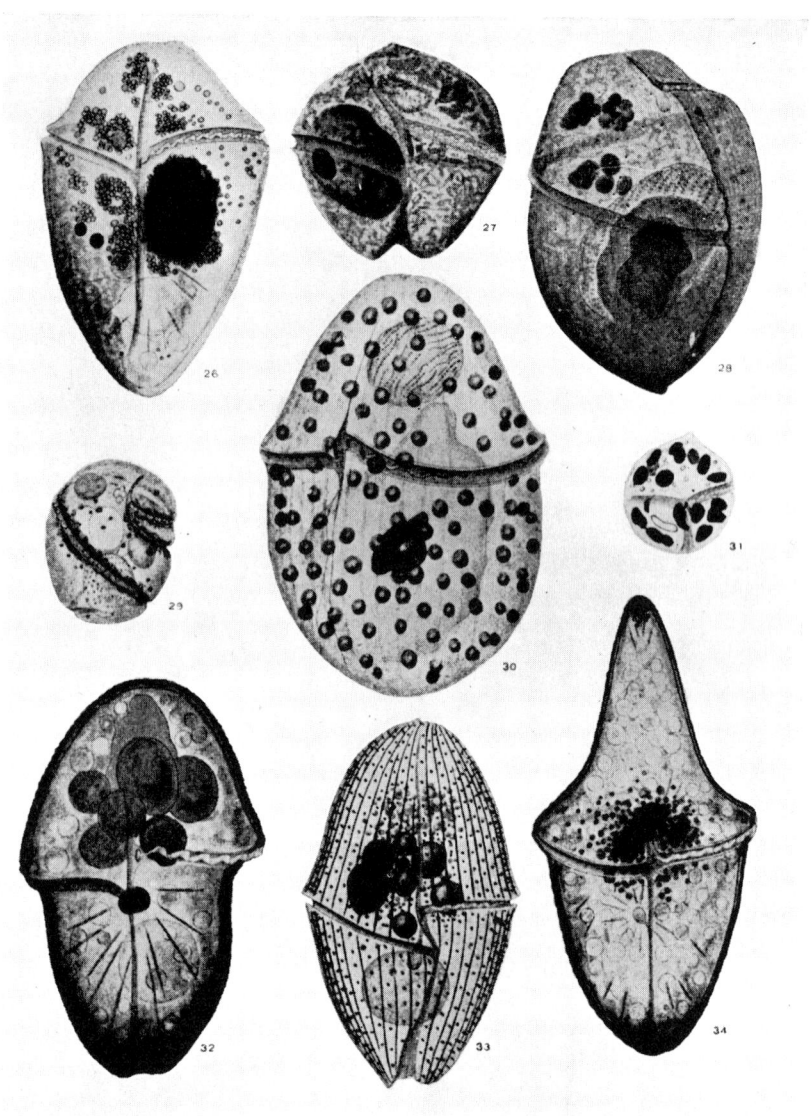

Fig. 24 Various types of luminous unarmored dinoflagellates (mostly *Gymnodinium*). The size varies from 16μ to 80μ in diameter. After Kofoid and Swezy.

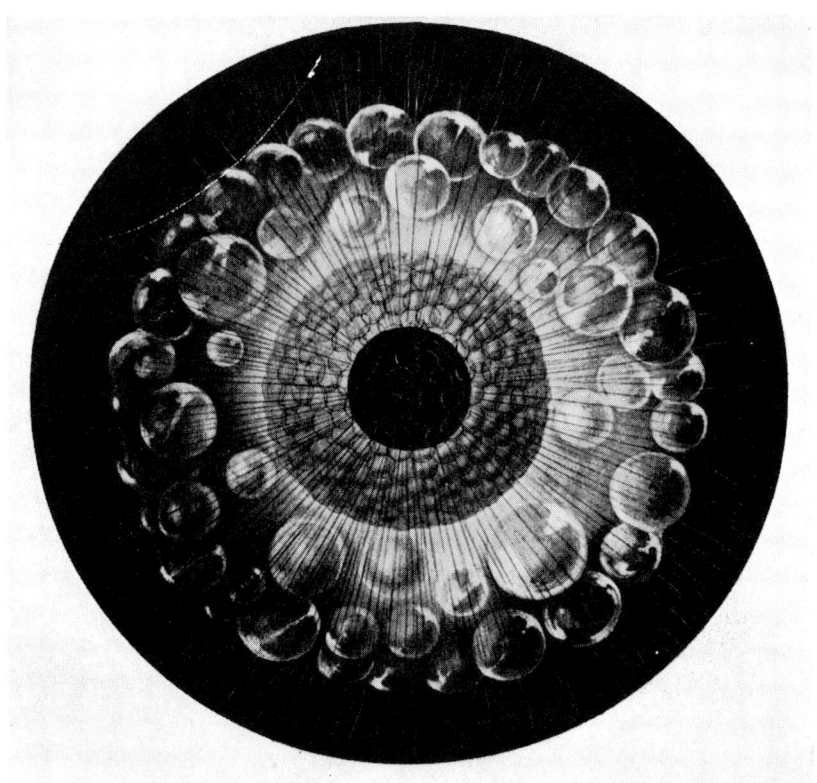

FIG. 25 A radiolarian, *Thalassicolla nucleata*, showing the central capsule and outer zone. After Huth.

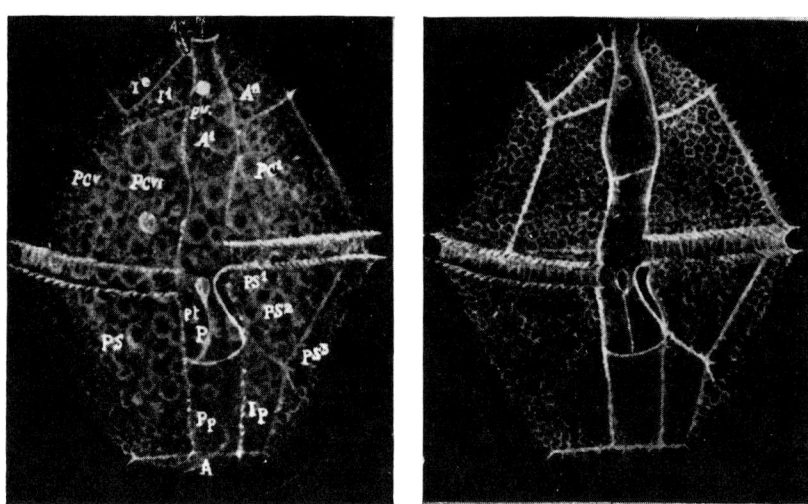

FIG. 26 The armored dinoflagellate, *Gonyaulax polyhedra*, which often increases in sufficient number to pollute waters of the Pacific Coast. These appear red by day and shine like fire by night. After Buhigas.

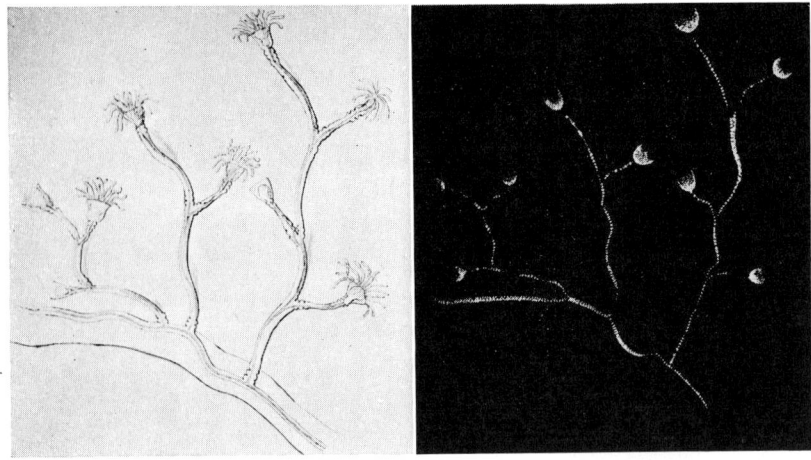

Fig. 27   Luminous hydroids by day (left) and at night (right), after Panceri.

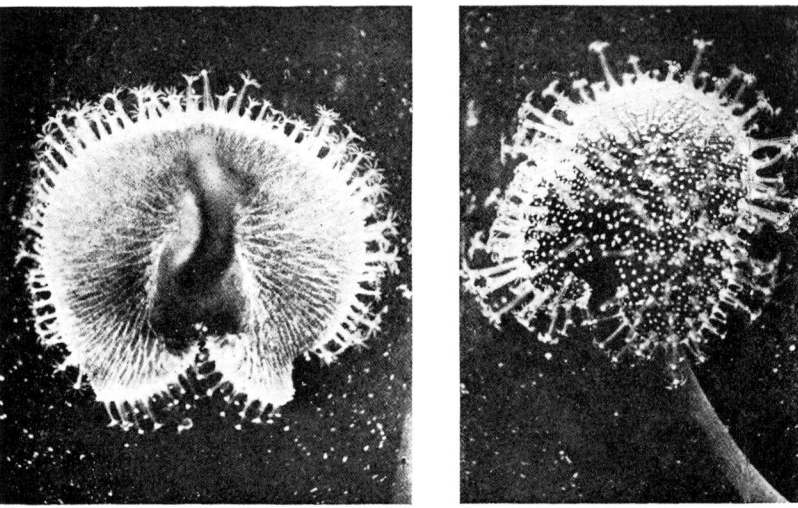

Fig. 28   The sea pansy, *Renilla*, a pennatulid. Note the kidney-like shape of the colony, consisting of individual polyps on a stalk (right). The individuals are so intimately connected by nerves that a touch in one region will cause a wave of light to pass over the colony. After Parker.

Fig. 29 Sea pens, luminous colonies of animals growing in sand, often at great depths. Reproduced by kindness of Ulric Dahlgren, from a drawing by Bruce Horsfall.

Fig. 30 *Mnemiopsis leidyi*, a common ctenophore at Woods Hole, Mass. The canals under the eight rows of swimming plates (like steps) become brilliant greenish luminescent when the animal is touched at night. From Buchsbaum's *Animals without Backbones*, by courtesy of University of Chicago Press.

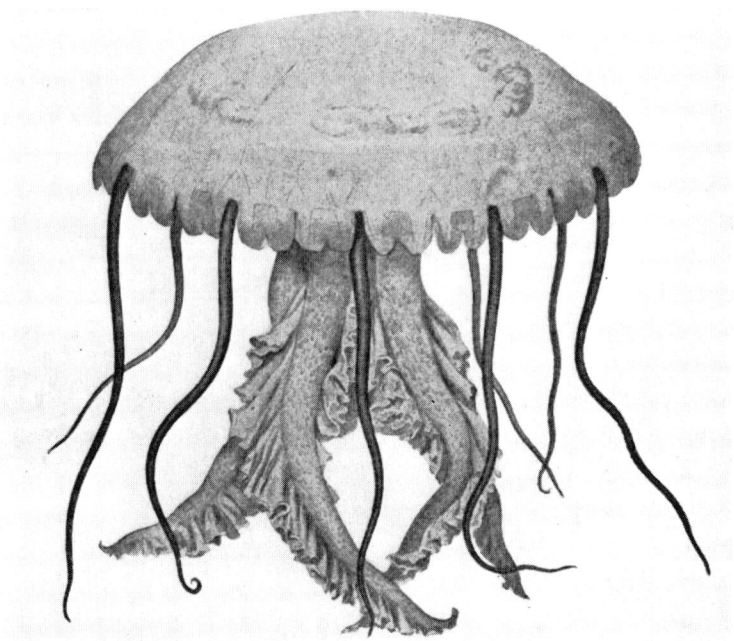

Fig. 31 *Pelagia noctiluca*, a luminous jellyfish common in the Mediterranean, whose light was recorded by Pliny. After Steuer.

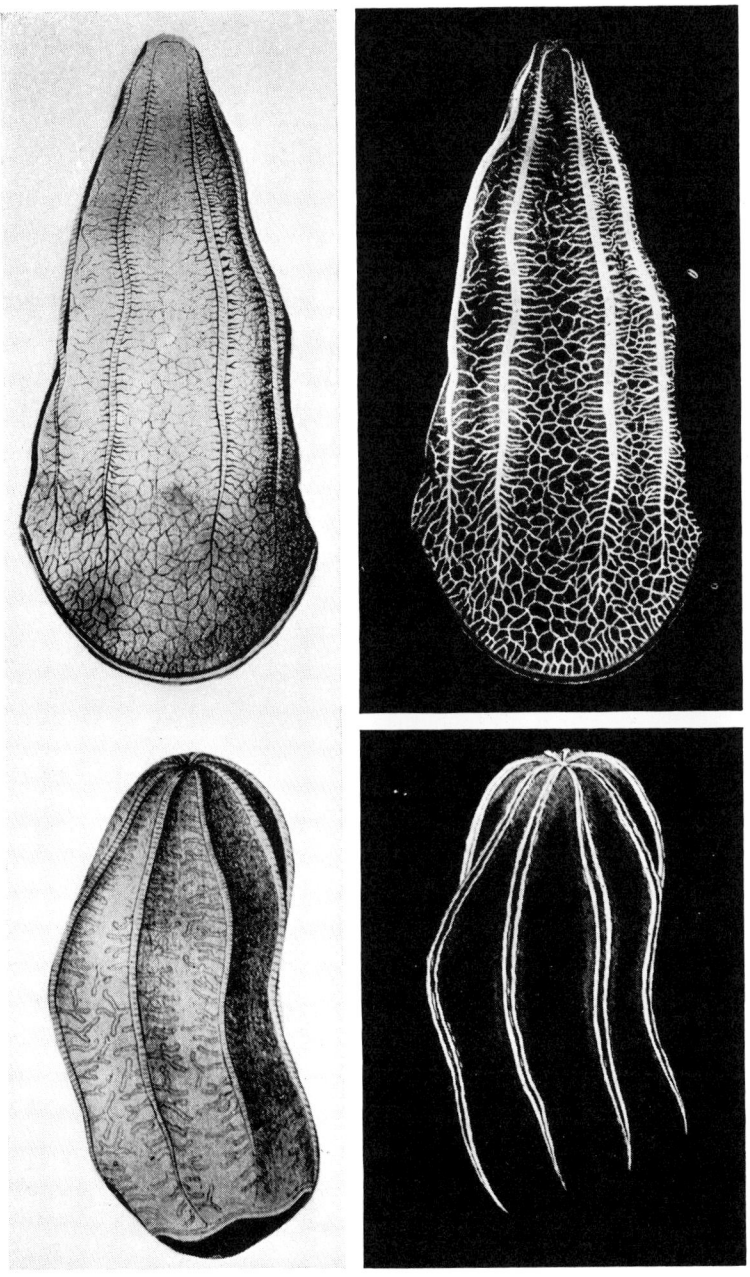

Fig. 32 Ctenophores or comb-jellies, by day (left) and at night (right). Note the delicate tracery of luminescent canals. After Panceri.

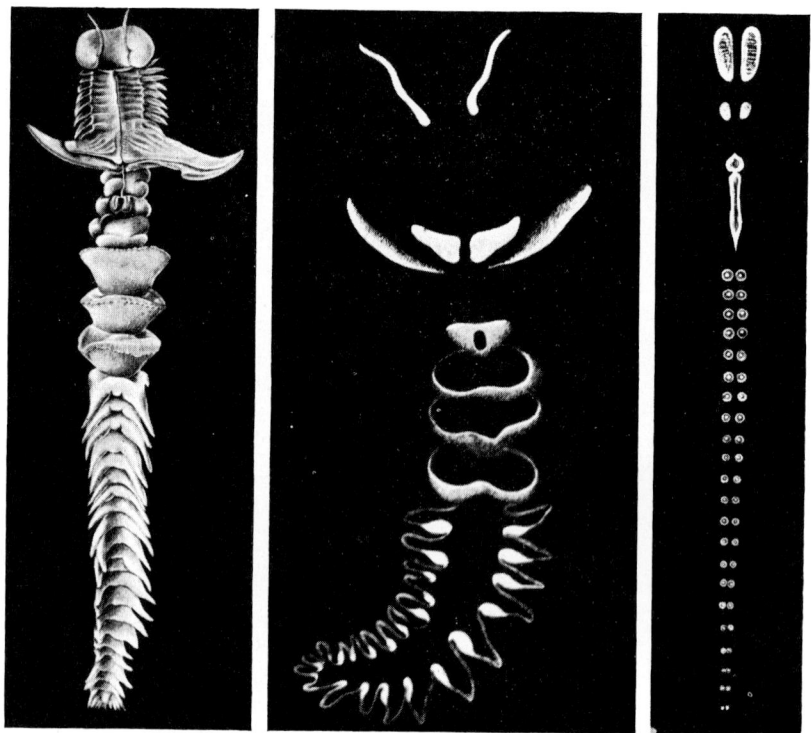

Fig. 33 *Chaetopterus* by day, after Trojan (left); at night after Panceri (center); *Mesochaetopterus* at night (right), after Fujiwara.

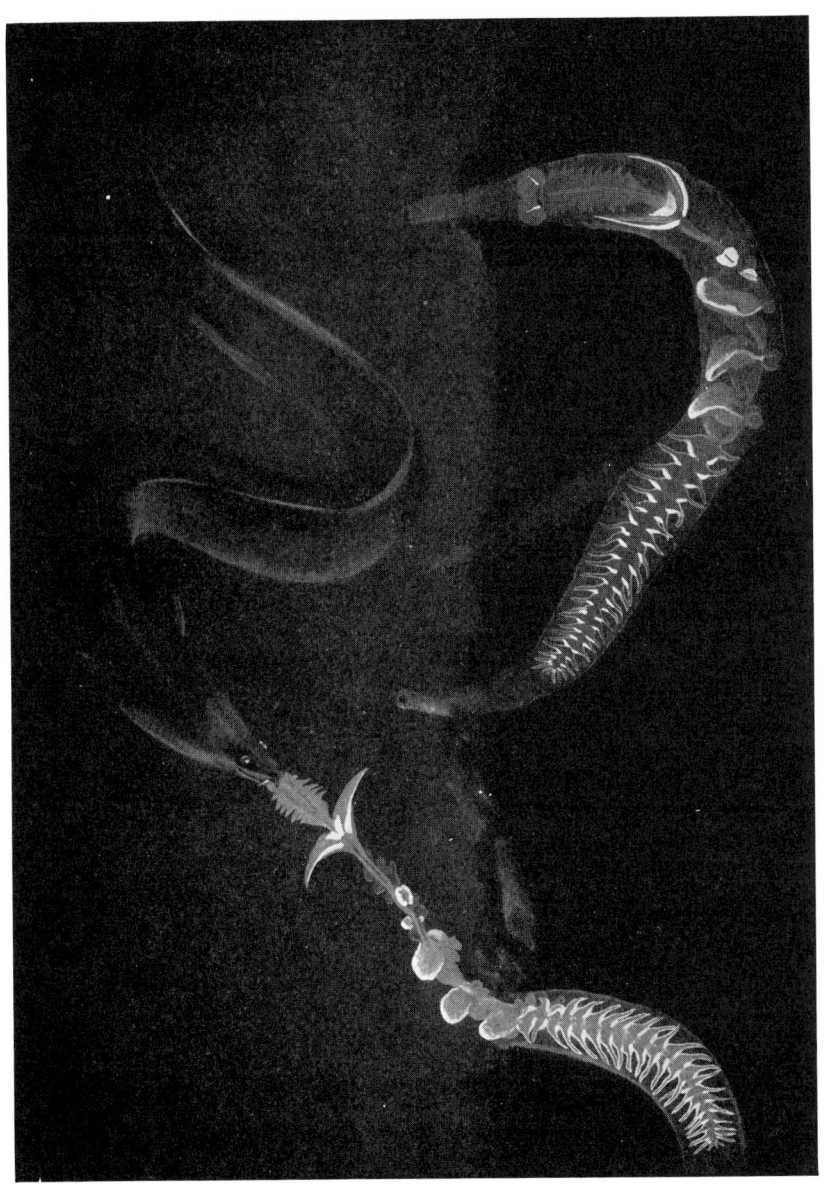

FIG. 34 Chaetopterus pulled out of its tube by an eel. Reproduced by kindness of Ulric Dahlgren from a drawing by Bruce Horsfall.

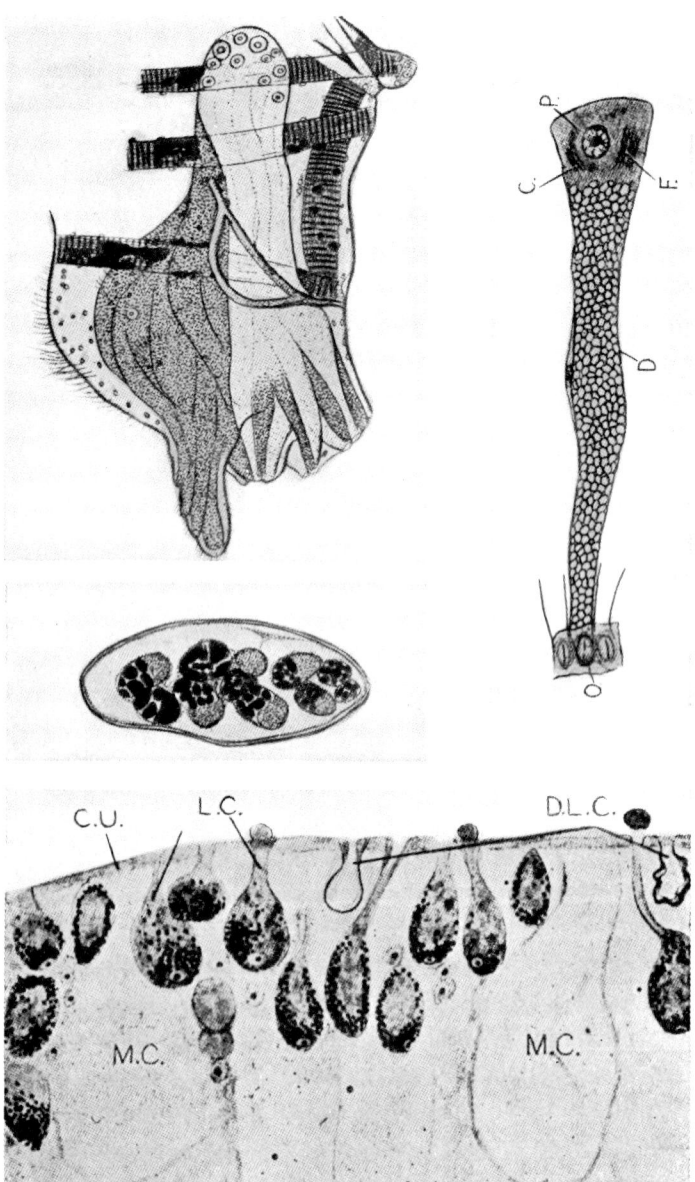

Fig. 35 A longitudinal section (above left) and a transverse section (center left) of the luminous gland of *Cypridina*, showing two types of gland cells which secrete luciferin and luciferase. After Yatsu. On upper right, a single gland cell of *Cypridina* full of granules, D; O, opening of gland cell to sea water; F, fibrillae; C, cytoplasm; P, nucleus. After Dahlgren.

Fig. 36 (Below) Transverse section of the luminous layer of *Chaetopterus*, showing luminous cells (L.C.) full of granules; discharged luminous cells (D.L.C.) and mucous cells (M.C.); C.U., cuticle. After Dahlgren.

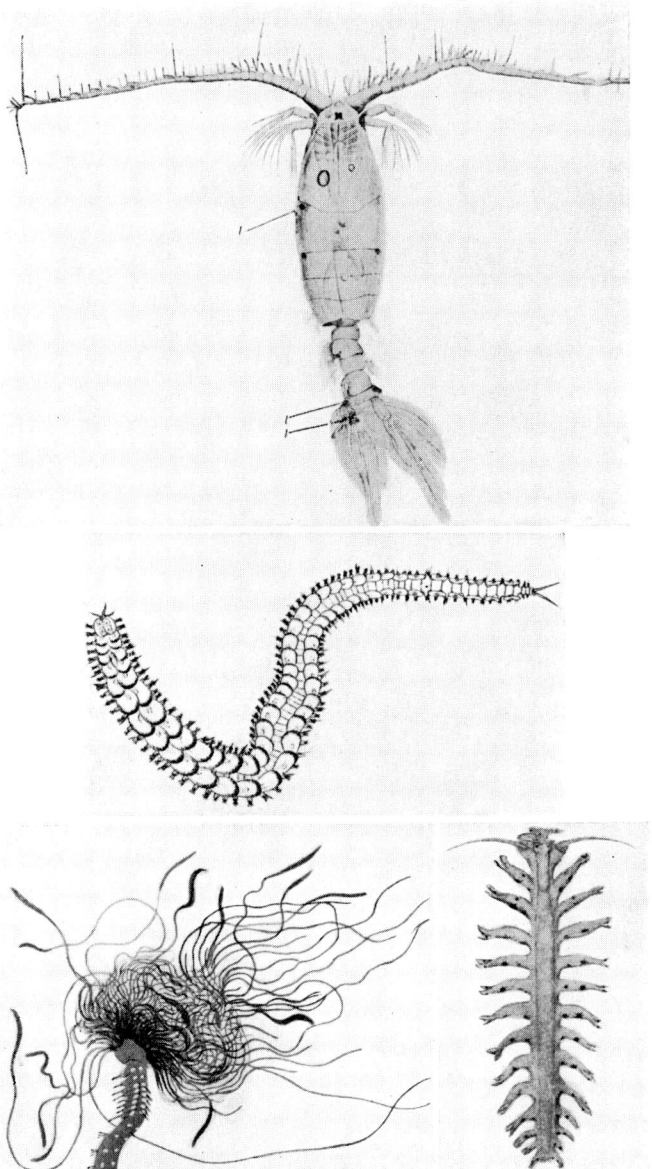

Fig. 37 (Below, left) A luminous cirratulid worm, with its mass of tentacles, such as was early observed living on oyster shells. After Panceri. Below, right. *Tomopteris rolasi*. After von Greef.

Fig. 38 (Center) The luminous polynoëd worm, *Acholoë astericola*. After Kutschera.

Fig. 39 (Above) A luminous copepod (*Pleuromma abdominale*) with light organs (1) on body and tail. After Giesbrecht.

FIG. 40 A polynoid worm attacked by a crab. The front portion ceases to luminesce and crawls away to regenerate a new tail, while the rear half, which cannot regenerate, is brightly luminescent to attract the attention of the crab. The light is used as a sacrifice lure. Reproduced by kindness of Ulric Dahlgren, from a drawing by Bruce Horsfall.

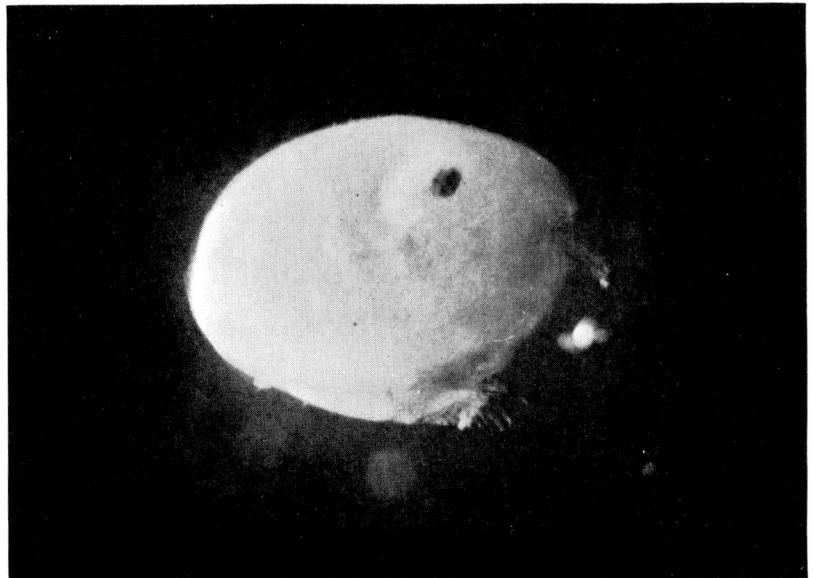

Fig. 41   *Cypridina hilgendorfii*. Above, photograph of a much magnified single animal showing black eye spot, tips of the legs projecting from under shell, and protuberance at right from which the luminescent secretion is ejected. Below, dried *Cypridina*, natural size, such as are used for extraction of the luminous substances, luciferin and luciferase. Original photographs.

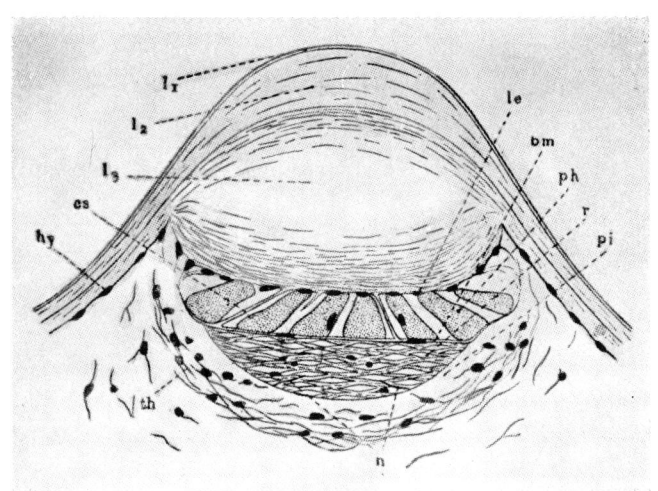

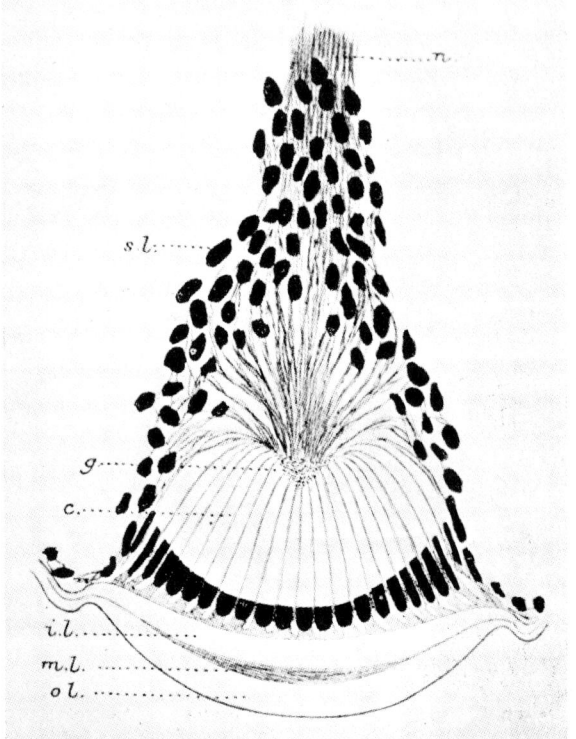

Fig. 42 Photophores of shrimp to show the lens-like thickening of chitin (l) and the photogenic cells behind it. Above, *Sergestes prehensilis*, after Terao. Below, *Acanthephyra debilis*, after Kemp.

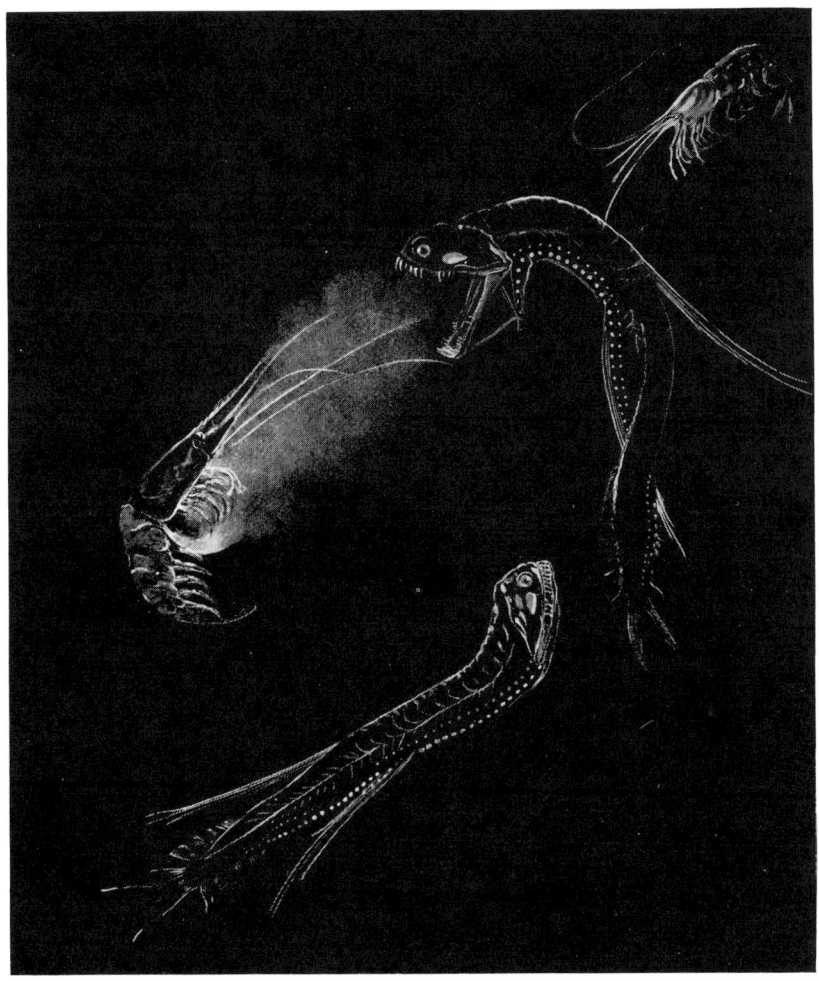

FIG. 43 A deep-sea shrimp, Acanthephyra purpurea, secreting from its luminous gland during a battle with the fish, Photostomias guernei. Reproduced by special permission from the National Geographic Society, after a painting by E. J. Geske.

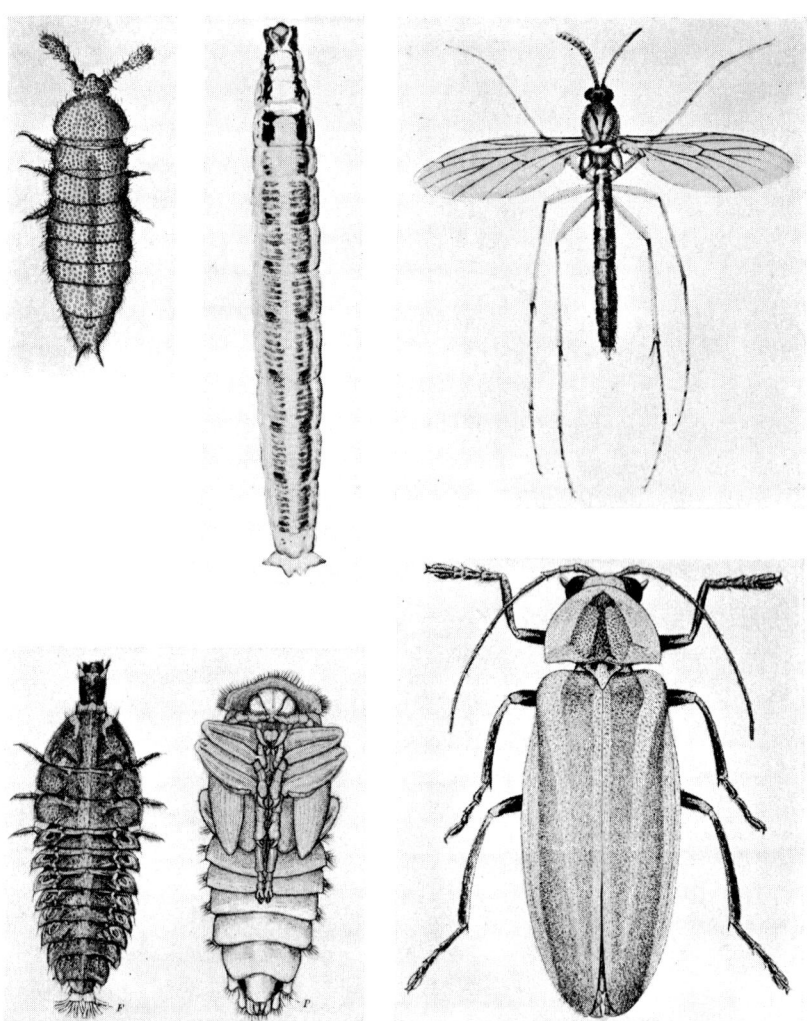

Fig. 44 Some luminous insects. Upper left, a collembolan, *Lipura*, after Henneguy. Upper right, non-luminous adult and upper middle, luminous larva of a fungus gnat, *Ceratoplanus testaceus*, after Stammer. Below, larva, pupa and adult of firefly, *Photuris pennsylvanica*. After Williams.

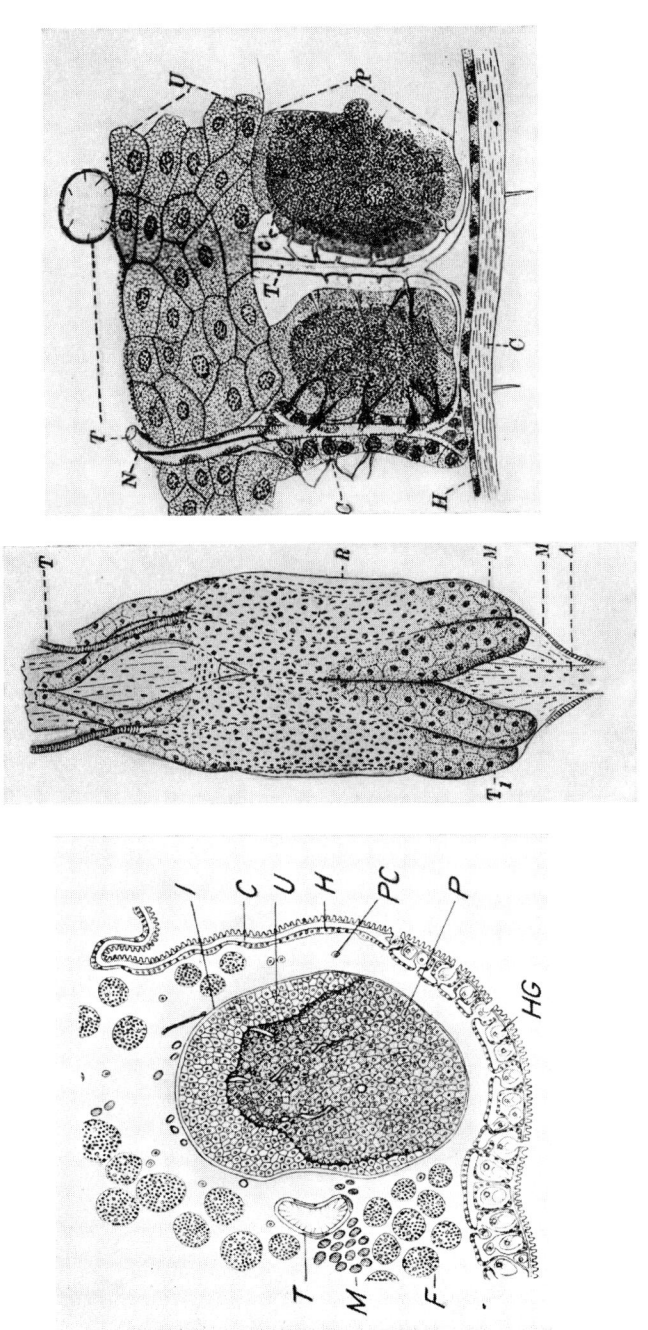

Fig. 45. (Left) Transverse section of the larval light organ of a Japanese glowworm, *Pyrocoelia rufa*, showing the photogenic cells, P, and the urate layer, U. Few tracheae penetrate the organ. After Okada.

Fig. 46. (Center) The light organ at the ends of the Malpighian tubules of the New Zealand glowworm, larva of a fly, *Boletophila*. After Wheeler and Williams.

Fig. 47. (Right) Transverse section of the light organ of an adult firefly, *Photinus consanguineus*, showing the urate layer, U; the photogenic cells, P; the tracheal trunk, T, with nerve, N, and the tracheal end cells, EC; H, the hypodermis; C, the cuticle. After Williams.

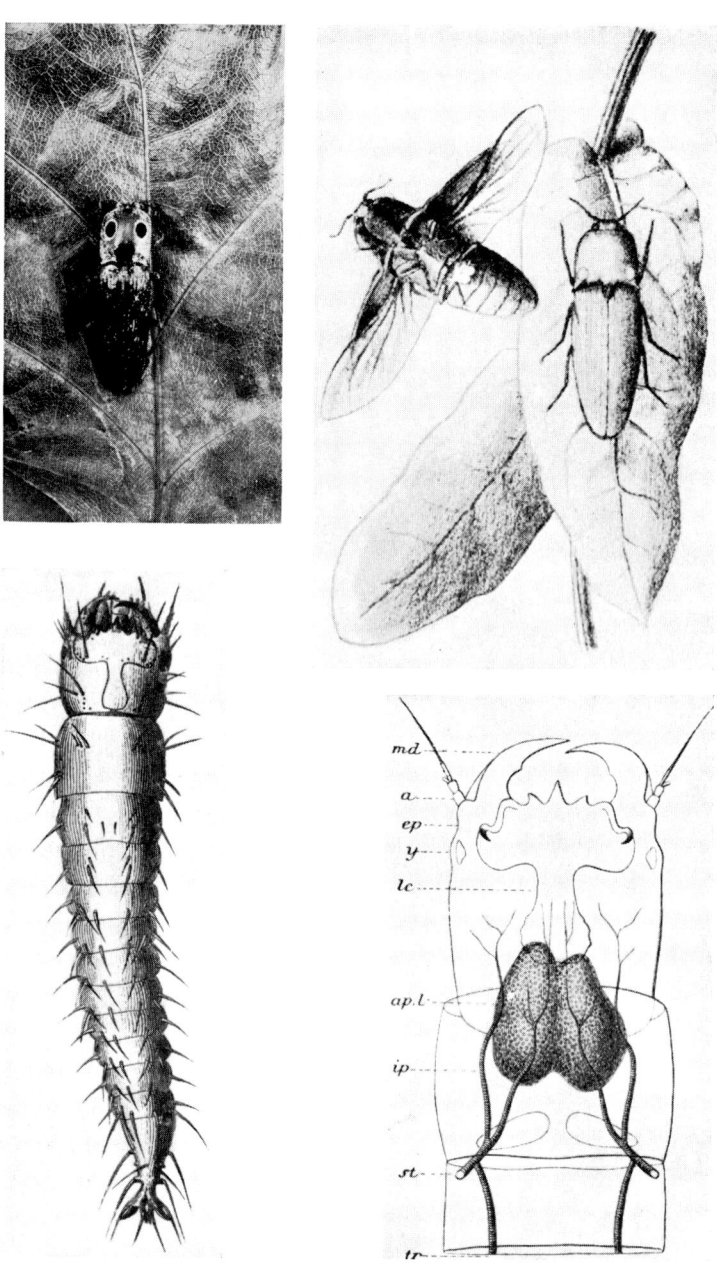

Fig. 48 At right top, the famous West Indian Firefly or Cucujo, *Pyrophorus noctilucus*, an elaterid beetle with three luminous organs, two on rear of prothorax, one on under side of abdomen, the latter only visible when flying; after Mangold. Upper left, the related nonluminous "death watch" has two black spots on prothorax. Below are shown a newly hatched larva of Pyrophorus (left) with its luminous organ (right) near head; after Dubois.

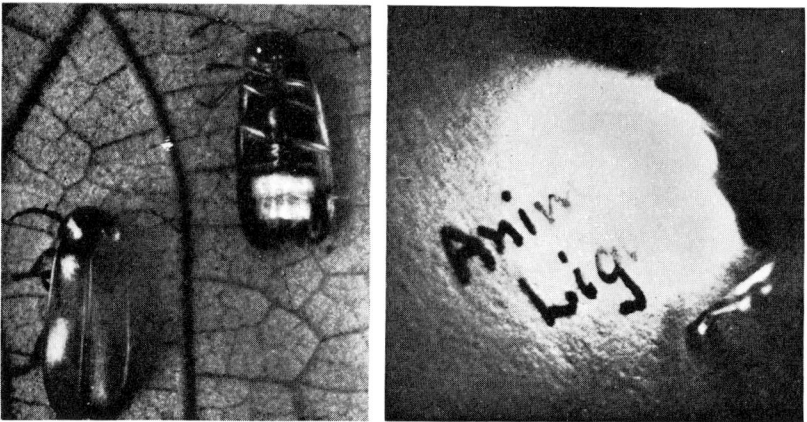

Fig. 49 The female beetle, *Phengodes*, a rare luminous form, found only twice by the author in thirty years of collecting. Photographs show the beetle crawling (above) and coiled, by day (left), and photographed by its own light, at night (right). Original photographs.

Fig. 50 A common firefly on a leaf and a photograph of its light at night. Photo by C. Clarke from Buchsbaum's *Animals without Backbones*, by courtesy of the Chicago University Press.

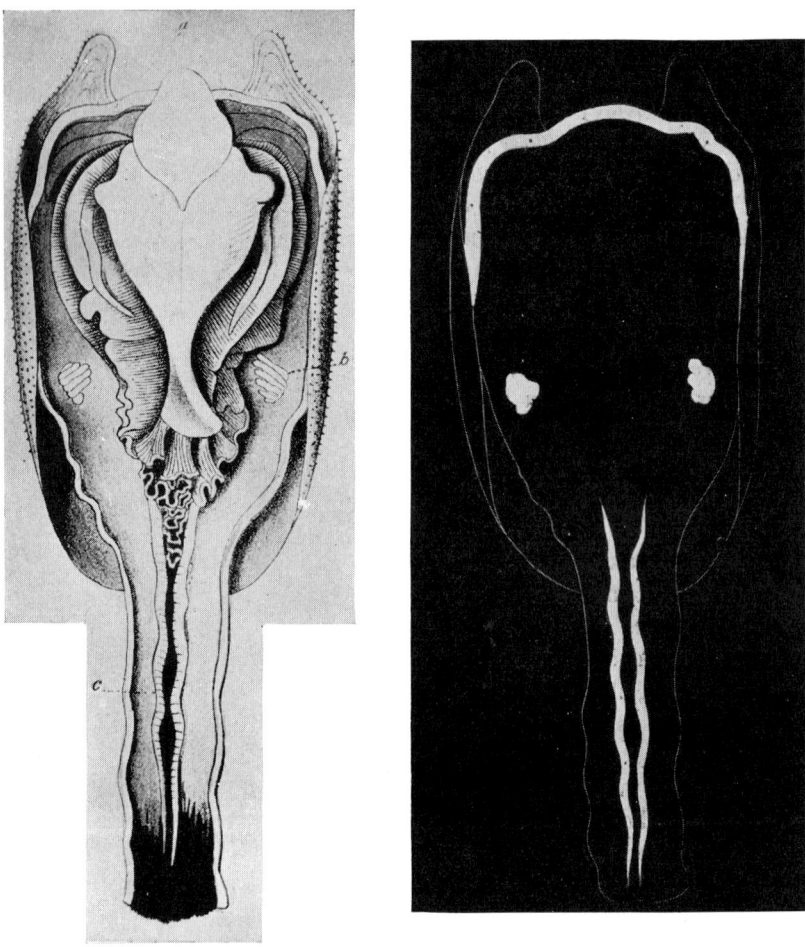

Fig. 51 The boring marine mussel, *Pholas dactylus*, by day (left) and by night (right) to show the five luminous regions; after Panceri.

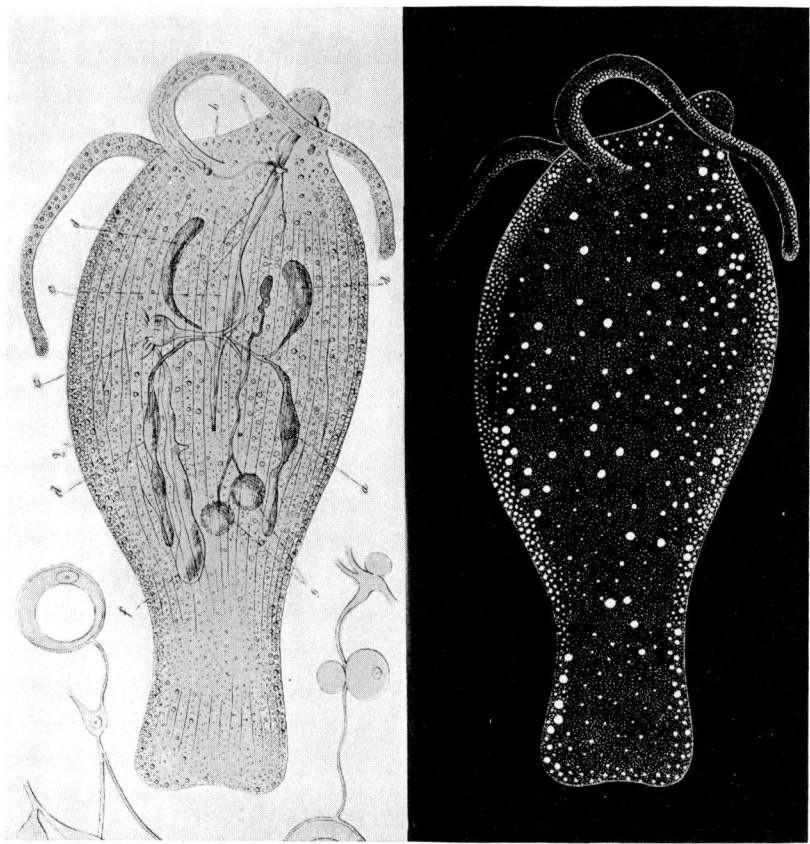

Fig. 52 The pelagic shell-less snail, *Phillirrhoë bucephala*, by day (left) and at night (right) showing luminescent spots; after Panceri.

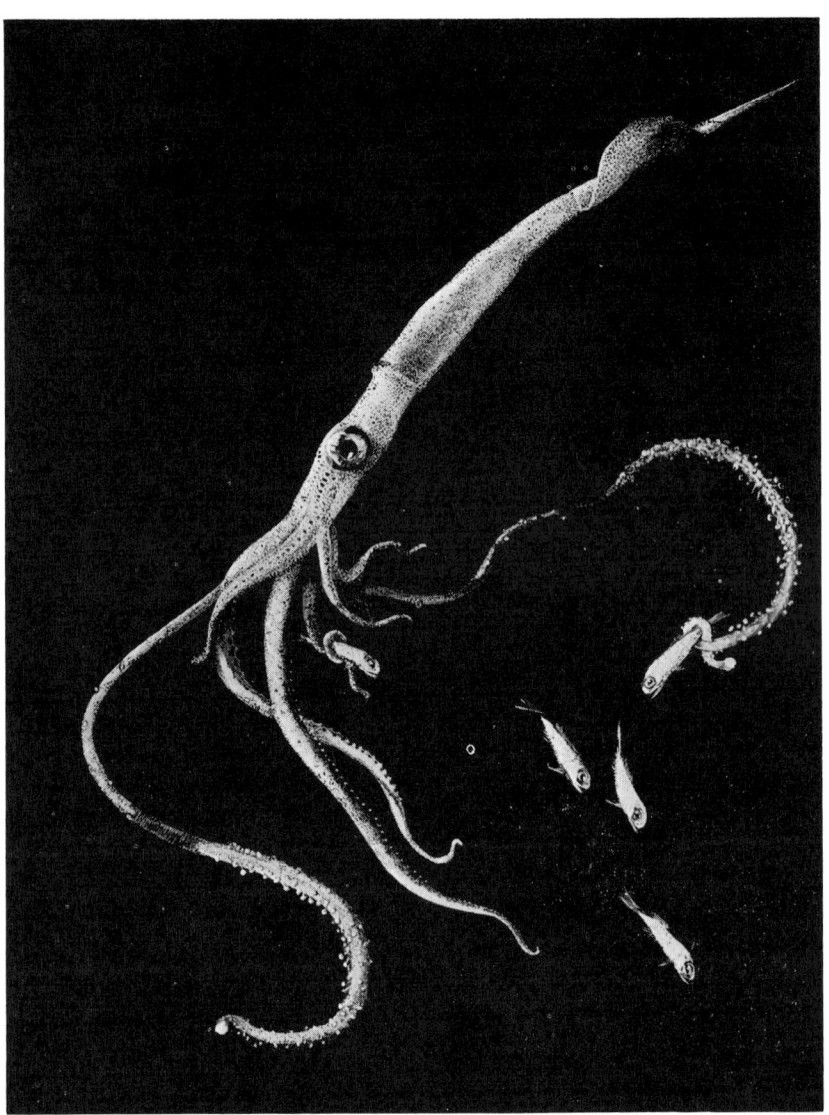

Fig. 53 A large luminous deep-sea squid catching lantern-fish. Reproduced by kindness of Wm. Beebe, after a painting by Else Bostelmann.

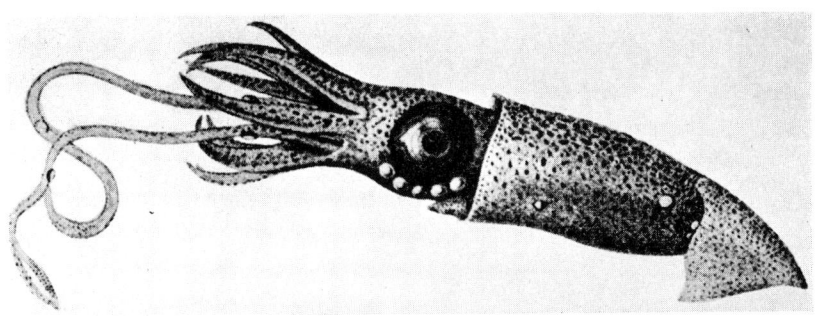

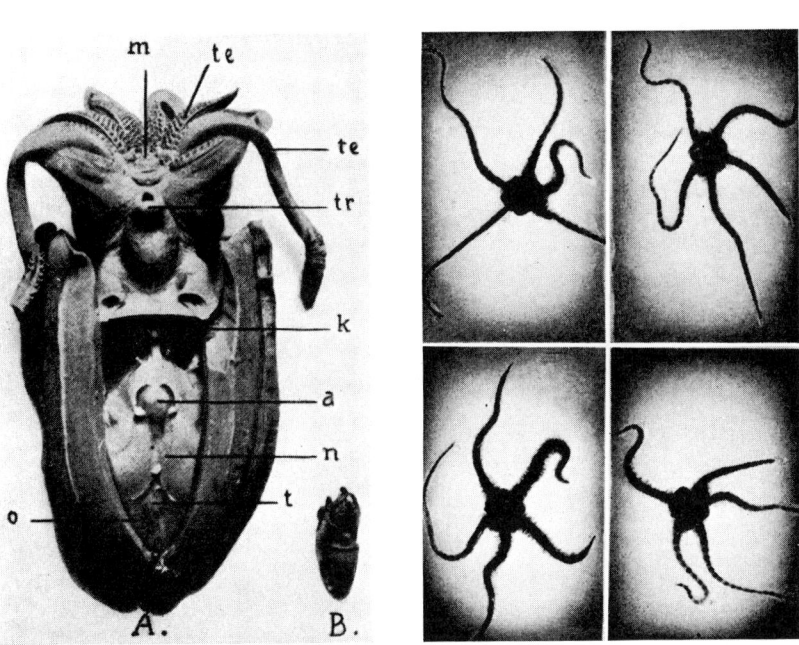

Fig. 54 (Above) A deep-sea squid, *Lycoteuthis* (*Thaumatolampas*) *diadema*, showing five large light organs on under side of eyeball and scattered organs over body; after Chun.

Fig. 55 (Left) *Sepia officinalis* (large) and *Rondeletia minor* (small) frequently containing harmless luminous bacteria while living. After Meissner.

Fig. 56 (Right) Snake-stars or brittle stars whose successive snake like movements are shown in the four photos. Some of the species are luminous. From Buchsbaum's *Animals without Backbones*, reproduced by courtesy of the Chicago University Press.

[293]

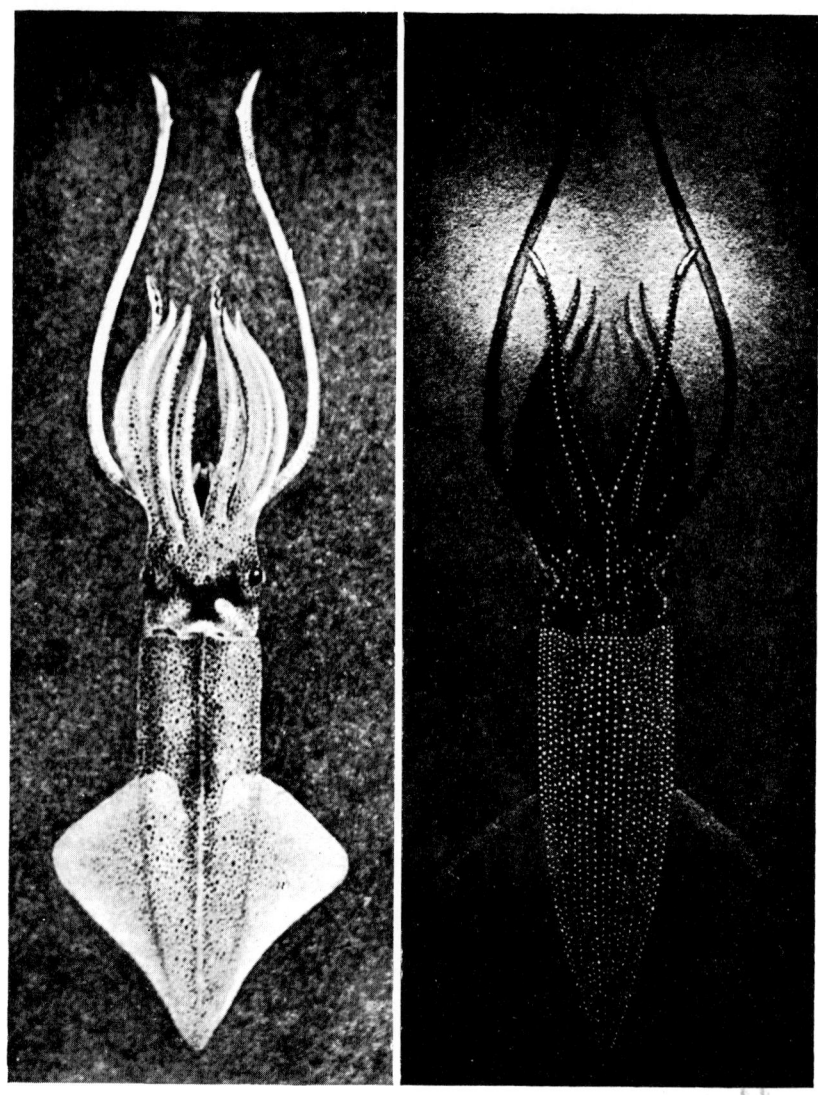

Fig. 57 The Japanese Firefly Squid, *Watasenia scintillans*, by day (left) and at night (right). Note three bright organs on short pair of arms, five on eyeball and many scattered luminous dots on body. After Sasaki.

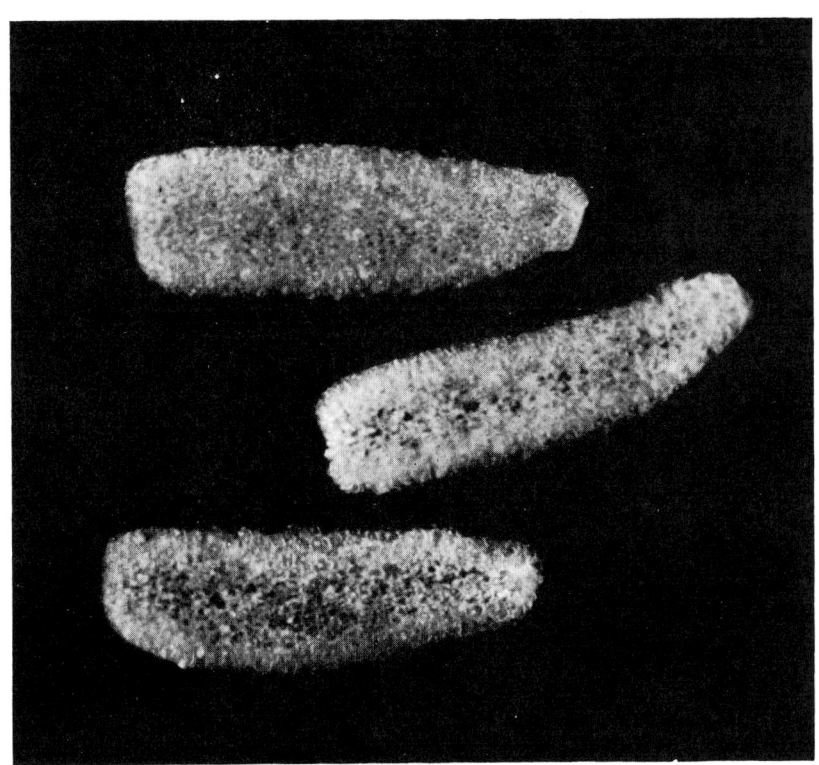
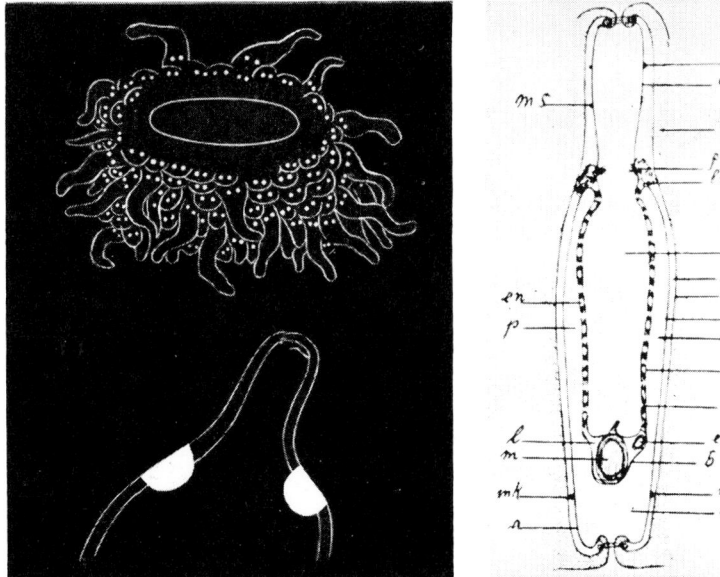

Fig. 59 Original photograph of three *Pyrosoma*, floating colonies of animals, each with two groups of luminous cells on the shoulder, shown diagrammatically in a cross section of the colony and of the animal (below); after Panceri.

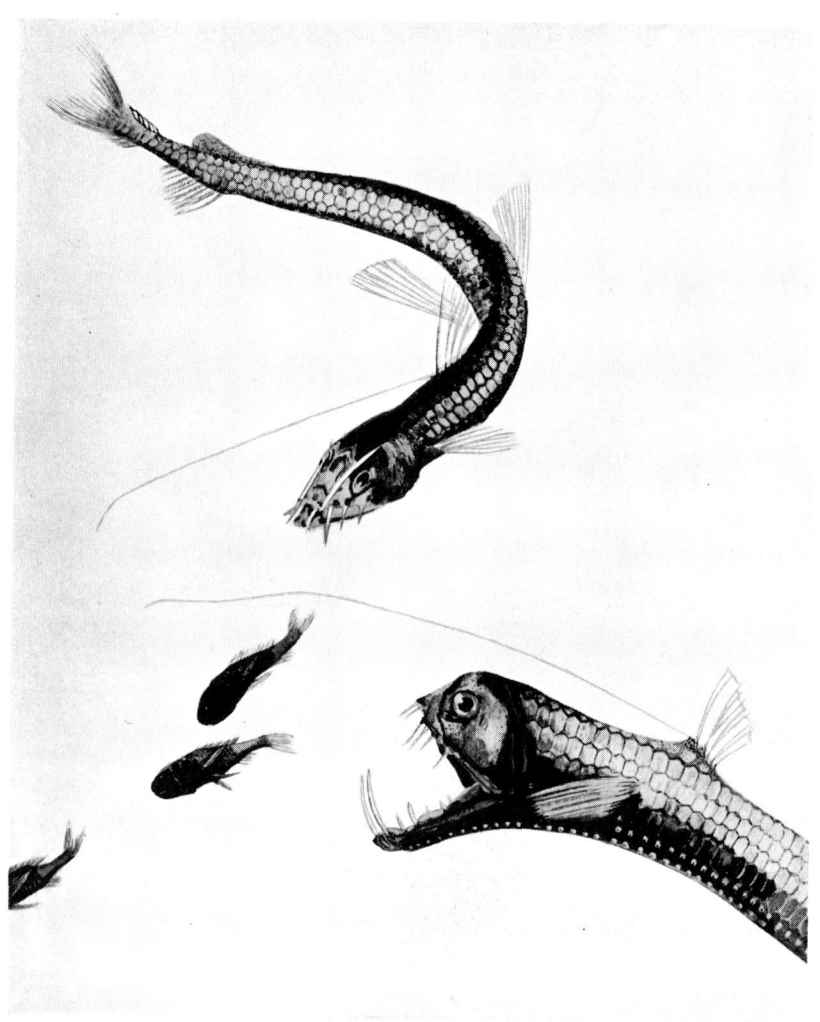

Fig. 60 The deep-sea fish, *Chauliodus sloanei*, attacking smaller forms (*Melamphoes crassiseps*). Reproduced by kindness of Wm. Beebe, from a painting by Else Bostelmann.

FIG. 61 The hatchet fish, Argyropelecus. Reproduced by kindness of Ulric Dahlgren, from a drawing by Bruce Horsfall.

FIG. 62 A deep-sea fish and deep-sea squid. Reproduced by kindness of Ulric Dahlgren, from a drawing by Bruce Horsfall.

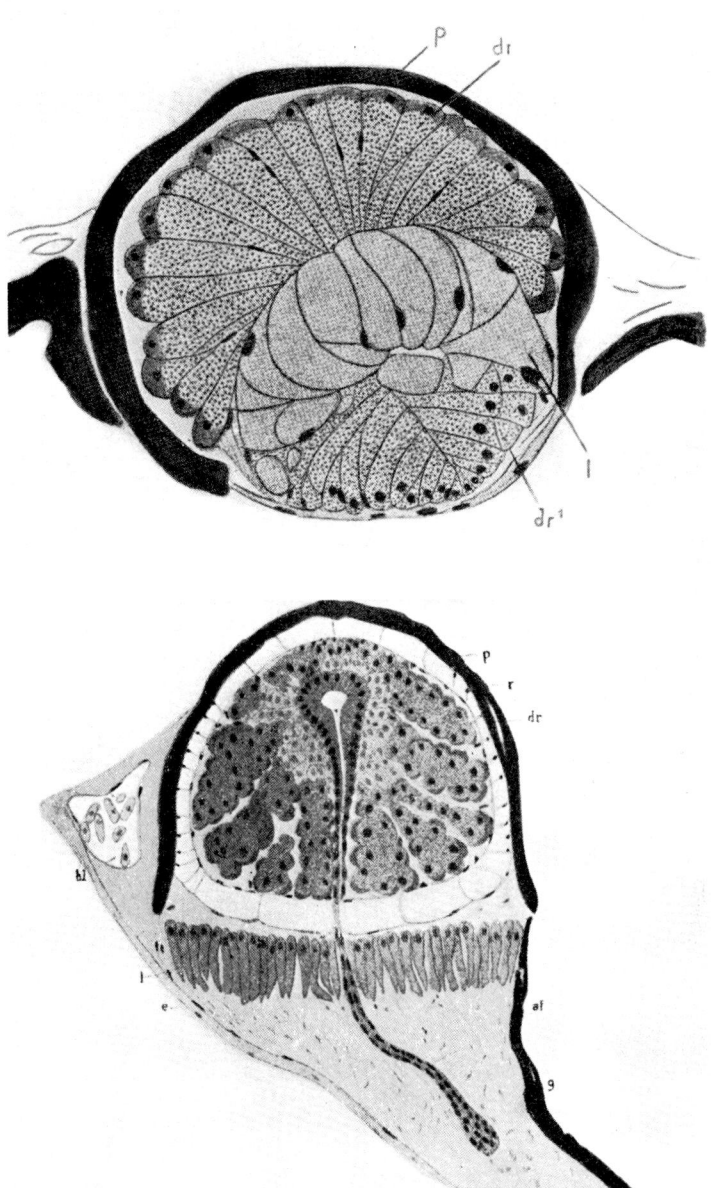

Fig. 63 Section of light organs of fish. Above, *Stomias*, without a duct. Below, *Cyclothone*, with a long duct turning to right. After Brauer.

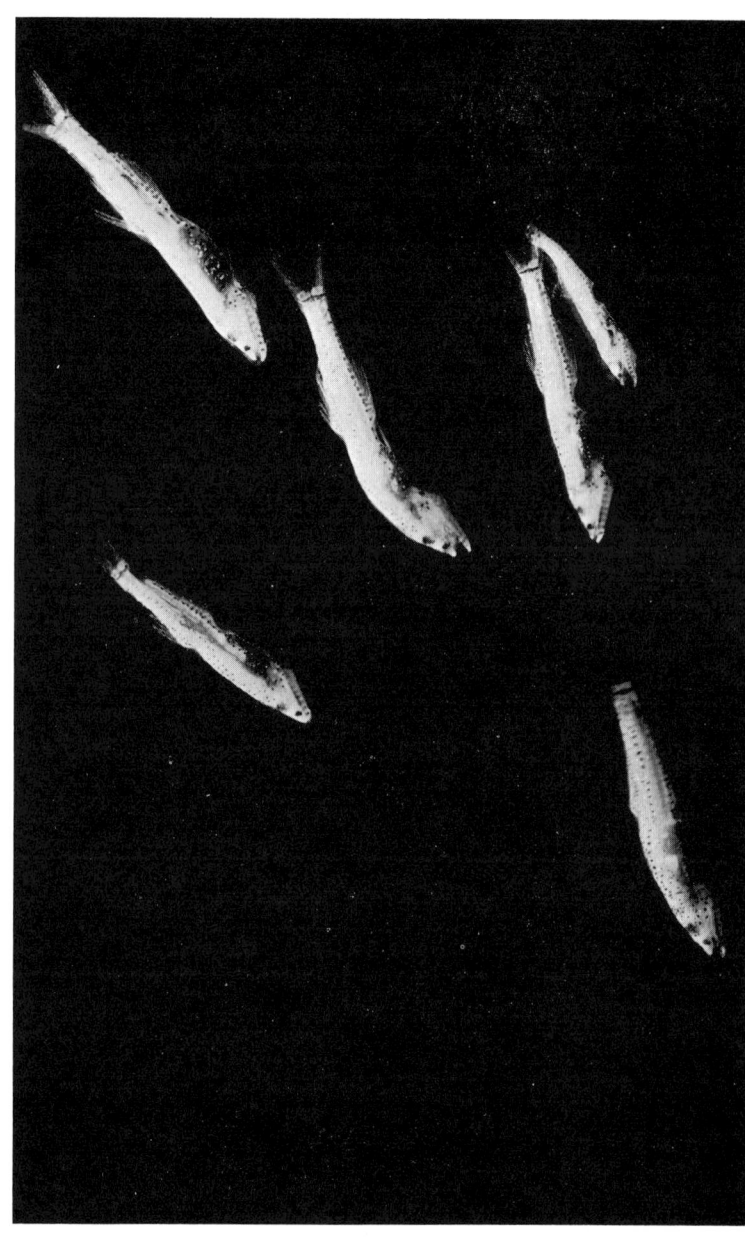

FIG. 64 Pale round-mouths, Cyclothone braueri. Reproduced by kindness of Wm. Beebe from a photograph of newly caught fish.

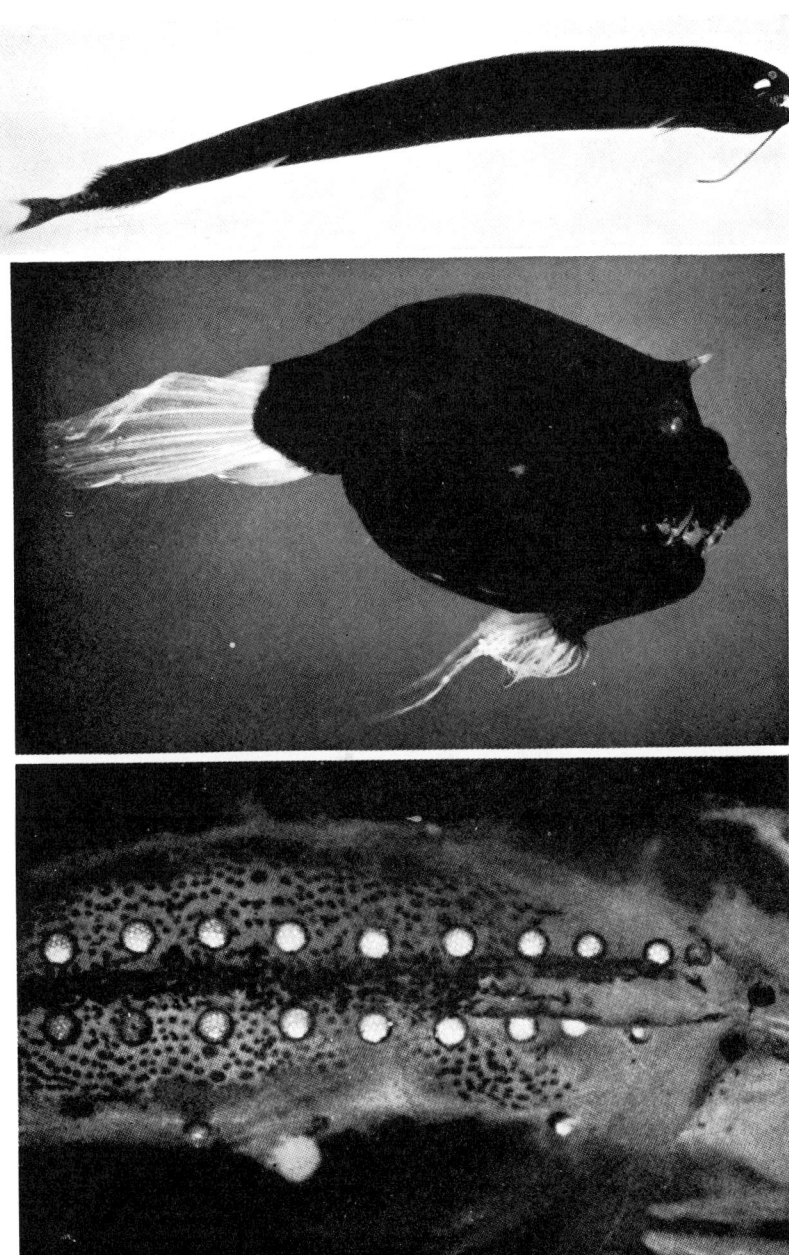

Fig. 65 (Above) *Echiostoma tanneri*; (Center) *Linophryne arborifera*; (Below) the double row of light organs of *Cyclothone*. All photos by Wm. Beebe.

Fig. 66 *Idiacanthus fasciola*. Three adult females (large) and two adult males (small) chasing shrimp. Reproduced by kindness of Wm. Beebe from a painting by Else Bostelmann.

Fig. 67 *Grammatostomias dentatus*. Note the enormously long barbel and prominent cheek light organ. Reproduced by kindness of Wm. Beebe, from a painting by Else Bostelmann.

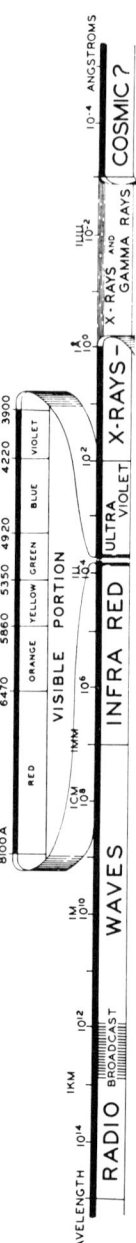

FIG. 68 Electromagnetic waves. Note the minute portion of the spectrum occupied by the visible region, which has been magnified and placed above the complete range of wave lengths. After Henshaw.

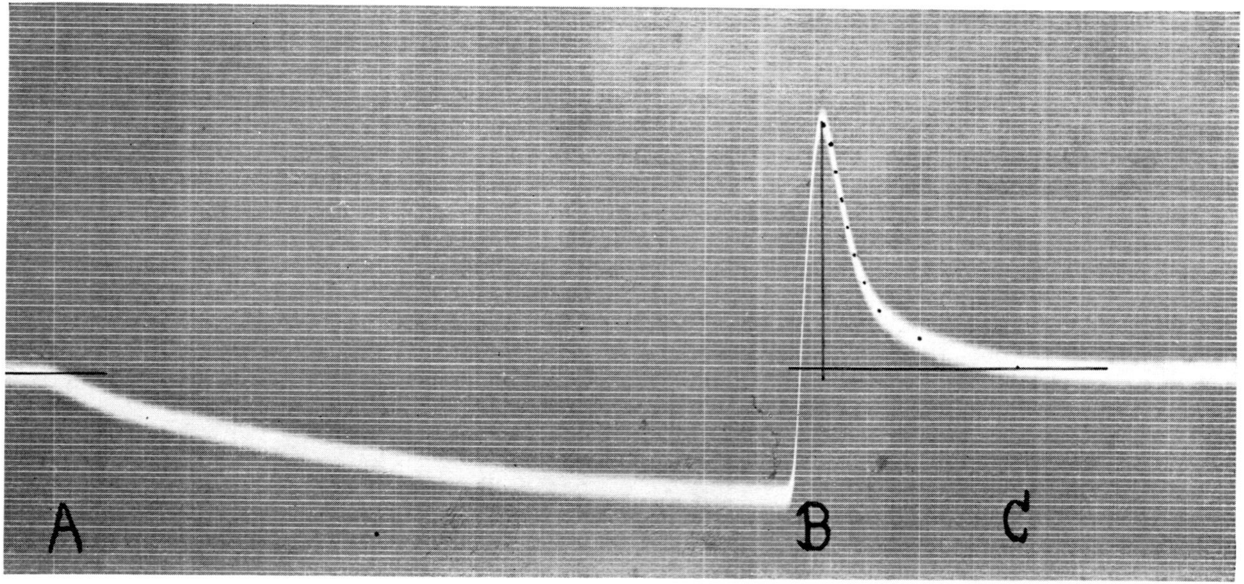

FIG. 69 A string galvanometer record of the light of luminous bacteria as they start to use up their own oxygen (A) and become completely dark (B). At B, more oxygen is admitted and a burst of luminescence ("excess luminescence") follows, with a return to the original steady light (C). Time on horizontal is 12 seconds from A to B and 4 seconds from B to C. Light intensity on vertical. Original record.

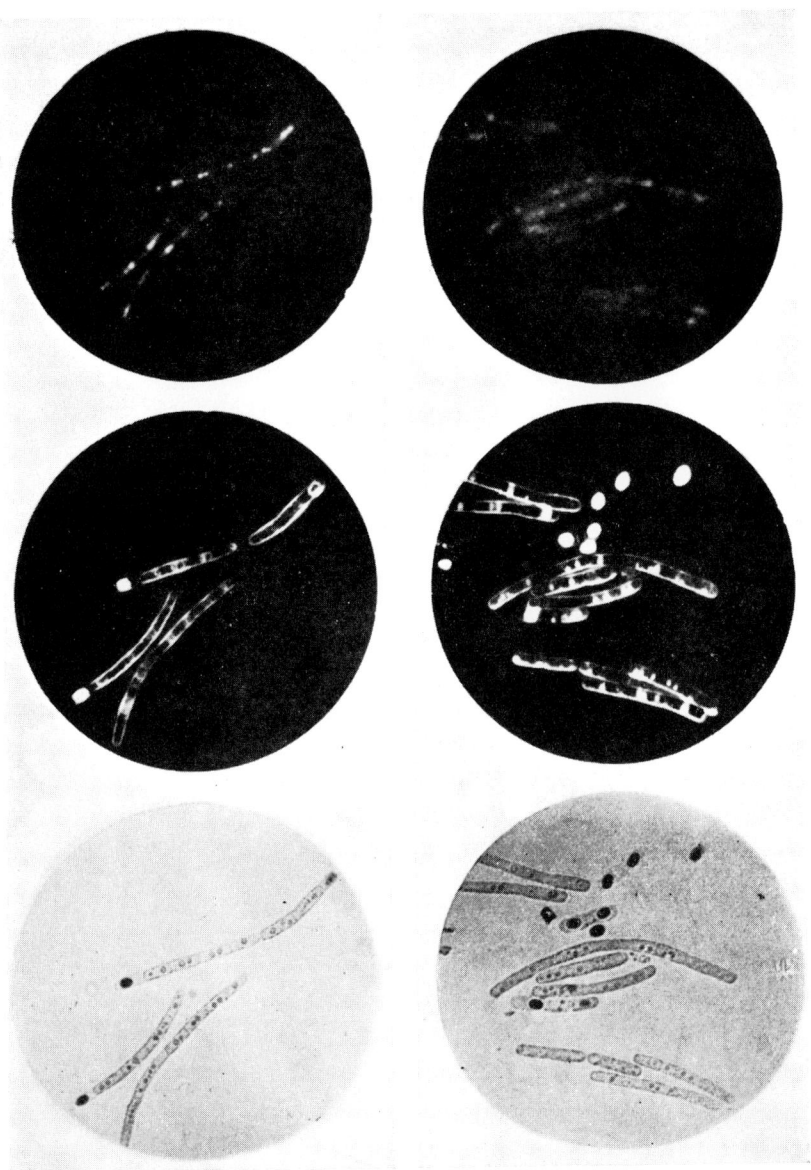

FIG. 72 Bacteria (*Bacillus anthracis*) magnified 1,000 times and photographed in (top) fluorescent light; (middle) dark field ultraviolet; (bottom) transmitted ultraviolet light. Note that the spores are non-fluorescent; after Barnard and Welch.

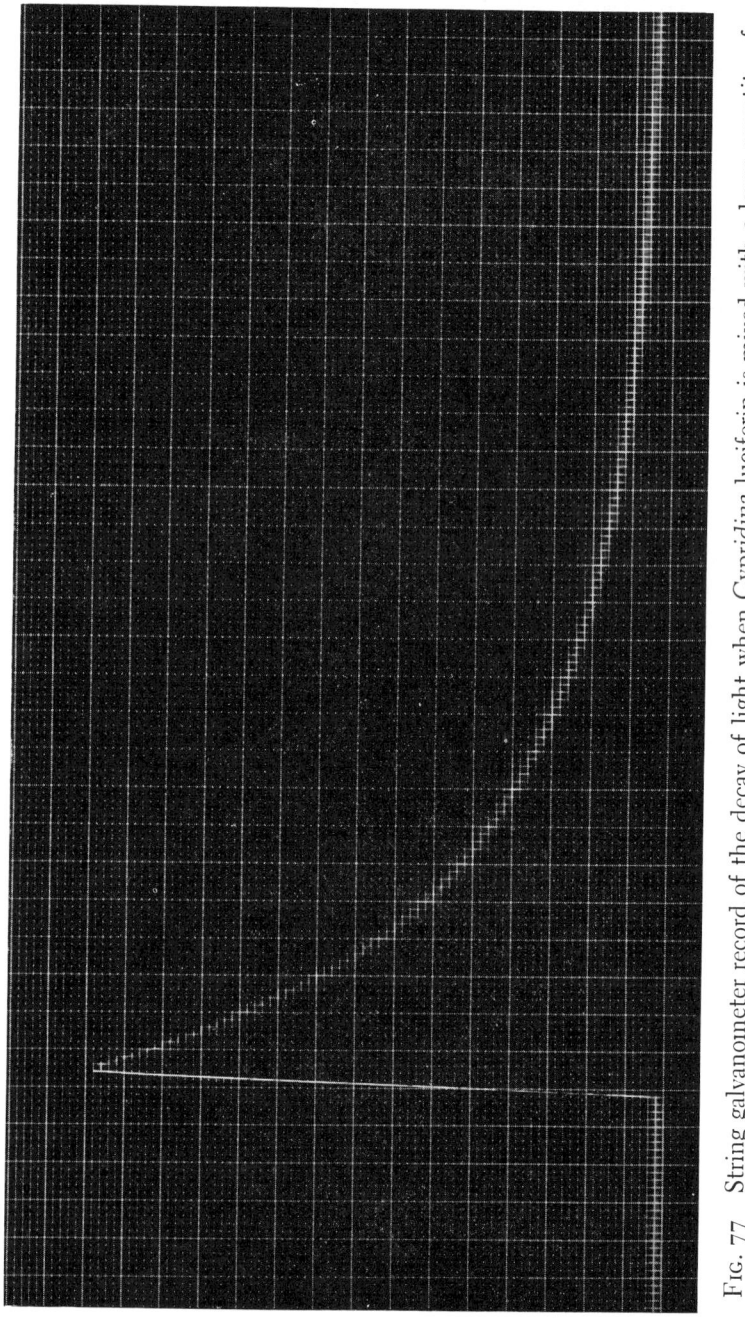

FIG. 77 String galvanometer record of the decay of light when Cypridina luciferin is mixed with a large quantity of luciferase. Rate of decay is very rapid, half complete in 0.7 seconds. Time in 0.2 second intervals along horizontal; luminescence intensity vertical. After Harvey and Snell.

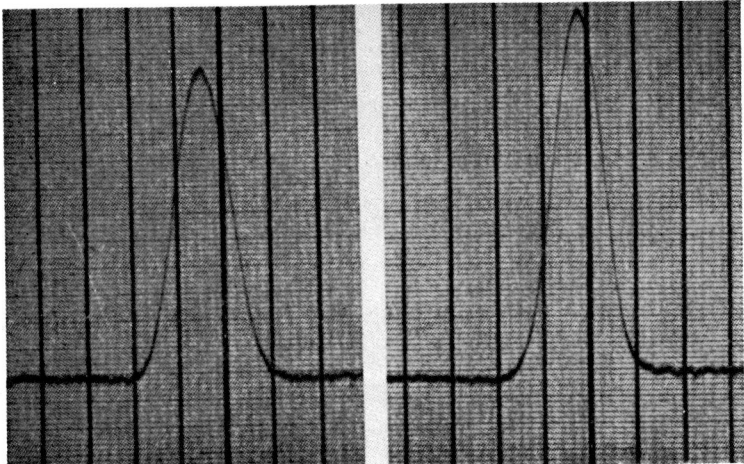

Fig. 78 String galvanometer records of two flashes of the firefly, *Photuris pennsylvanica*. Light intensity vertical; time in 0.04 second intervals horizontal. Note that each flash is symmetrical and lasts for 0.12 seconds. After Snell.

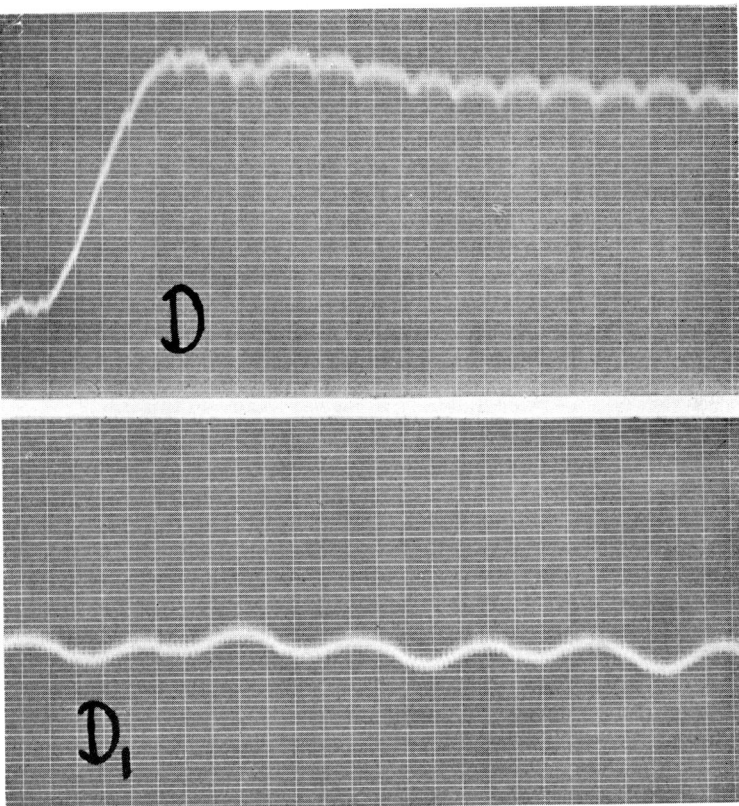

Fig. 79 String galvanometer record of the light of the elaterid beetle, *Pyrophorus*. Light intensity vertical; time in 0.1 second intervals horizontal. Fifteen seconds elapse between end of D and beginning of D_1. Note the slow development of light and rhythmic variation in intensity, 6 per second at first, slowing to 2.5 per second. Original record.

INDICES

AUTHOR INDEX

Adams, 94, 214
Agassiz, 50, 52
Akabane, 180
Albrecht, 117, 121
Alcock, 64-5, 82
Alenitzin, 67
Alexander, 73
Allman, 52, 66
Amberson, 149-52
Anderson, 61, 127, 132, 137, 138, 139, 144, 145, 152, 153, 187
Andrews, 100
Annandale, 68
Arago, 18
Aristotle, 6
Armstrong, 100
Arndt, 65
Audouin, 65
Azara, 21

Backman, 117
Bacon, 16, 68, 108
Baird, 46
Bajon, 11
Baker, 6, 11, 44
Baldouin, 99
Ballner, 171
Baly, 94
Bancroft, 96, 107, 113
Banks, 61
Barber, 69
Barnard, 104
Bartolin, 26, 68
Baster, 44
Beal, 26
Beccarius, 75, 100
Becquerel, 99, 100
Beebe, 64, 82, 83, 87
Beijerinck, 27, 28, 130, 156, 179, 180, 181
Bennett, 82

Benning, 67
Bernouilli, 105
Berry, 76, 79
Bhatnagar, 121
Billet, 13, 14, 29
Biornonius, 41
Bischof, 40
Blair, 68
Bleeker, 31
Bloch, 114
Blondlot, 223
Boddaert, 31
Boisduval, 29
Bongardt, 70, 167
Born, 76
Bose, 39
Bothe, 41
Bourzes, 7-8
Bowdoin, 10
Bowring, 72
Boyle, 6, 9, 26, 37-9, 40, 44, 96, 98, 99, 108, 123
Brade-Birks, 65
Brandis, 83
Brandt, 3, 46, 47, 99
Brauer, 63
Brewster, 103
Brockhausen, 65
Brodhurst, 65
Broeckmann, 40
Brown, 30, 64, 166
Brugiere, 55
Buchner, 15, 30, 81
Buck, 72-4, 166, 178
Buller, 40, 41
Burckhardt, 83
Burghause, 81
Burmeister, 69

Calvert, 121, 199
Canton, 4, 26, 99

INDEX

Carradori, 40, 124, 166
Cartesius, 9
Cascariolo, 98
Centnerswer, 114, 121
Chace, 63
Chambers, 106, 107, 172
Champion, 75
Chance, 154
Charpentier, 223
Chase, 139, 143
Chiarini, 83
Child, 94
Chun, 17, 63, 64, 76, 79, 200
Clarens, 126, 184, 188-90
Clarke, 63
Claus, 222
Coblentz, 146, 199, 201, 202, 207, 209
Cocco, 83
Cockayne, 104
Coolidge, 97, 110
Creighton, 167-9
Crie, 40
Crookes, 5
Crozier, 174
Cunningham, 63
Curie, 99, 101

Dahlgren, 34, 41, 46, 47, 51, 52, 57-60, 62-4, 70, 75, 83, 167, 168
Dankwortt, 104
Dartous de Mairan, 44
Darwin, 43, 48
Davy, xiv
de Coulon, 171, 183, 189
de la Voie, 57
Decker, 119
Deitrick, 139, 140
Delepine, 115
Derrien, 147
Derschau, 40
Dessaignes, 26, 99
Dewar, 100
Dewey, 218
Dicquemare, 45, 52

Diefenhorst, 121
Doepner, 199
Doflein, 62
Doudoroff, 193
Dougherty, 118
Downey, 121
Drew, 118, 119
Driessen, 21
Dubois, 24, 31, 35, 65, 66, 68, 75, 112, 114, 130-6, 140, 141, 145, 147, 166, 167, 198, 199, 201, 206, 217
Dufay, 105
Dufford, 108, 117, 121, 199, 214
Dushman, 216

Eaton, 30, 66
Eder, 115
Ehrenberg, 6, 46, 54
Eliot, 75
Ellis, 50
Elsholtz, 96
Emerson, 169
Emery, 70, 83
Enders, 57
Evans, 121, 220
Ewan, 114
Eymers, 121, 188, 189, 202, 203, 204, 209, 214
Eyring, 220

Fabre, 41
Fabricius ab Aquapendente, 26, 61
Falger, 60
Faraday, xiv
Fernandez de Oviedo, 65
Flaugergues, 55
Flosdorf, 107
Fischer, 26, 201
Fonda, 101
Forbes, 82
Forster, 75
Forsyth, 207
Forsythe, 165
Found, 216

INDEX

Fowler, 63
Frankland, 217
Franklin, 9-10
Frazer, 58
Frenzel, 106
Friedberger, 199
Fuchs, 68
Fuhrmann, 183
Fujiwara, 57

Gadeau de Kerville, 54
Gaertner, 40
Galeati, 75
Galileo, 98
Galloway, 58-9
Garman, 65
Garrison, 94, 113, 114
Gatti, 83
Gazagnaire, 65
Geipel, 70, 167
Geise, 104
Gentil, 11
Gernez, 109
Gerretsen, 141, 167, 168, 178, 179, 180, 183
Gesner, 49, 50
Ghave, 19
Giard, 14, 29
Giesbrecht, 62, 63
Giesel, 99
Giglioni, 50, 55, 75
Gilchrist, 56
Gimmerthal, 29
Giobert, 111
Gleu, 118, 119
Godeheu de Riville, 61
Goss, 115
Grant, 50, 104
Greef, 60
Greene, 83, 85, 165
Grew, 13
Grignard, 108, 117, 214
Grimm, 55
Grinaldi, 103
Griselini, 57

Gudden, 93
Gueguen, 40
Gueneau de Montbelliard, 71
Guinchant, 115
Gurwitsch, 221, 223
Guyton-Morveau, 21

Haase, 65, 69
Hablizl, 67
Haitinger, 104
Hall, 71
Hamperl, 104
Handrick, 83
Haneda, 33, 34, 84, 133
Hankel, 26
Hanna, 41
Hansen, 63
Harms, 34
Harris, 118, 214
Harvey, E. N., 14, 27, 29, 31, 33, 46, 47, 51, 58, 59, 64, 70, 71, 86, 87, 94, 104, 105, 106, 110, 115, 118, 124, 126, 127, 128, 131, 133-5, 137, 139, 140-8, 152-9, 162, 165, 168, 172, 175-9, 181, 184, 185, 189, 191, 192, 198-202, 213, 214
Harvey, E. B., 162-4, 171-3, 182-3, 189
Haber, 220
Hawkesbee, 105
Hayashi, 36
Hazen, 30, 66
Heber, 74
Heczko, 117
Heimstadt, 104
Heinemann, 68, 166, 178
Heinrich, 40
Helberger, 116
Heller, 6, 7, 26, 41
Henneberg, 67
Hepp, 67
Herfurth, 35
Hermbstadt, 26, 41
Herschel, 103

INDEX

Hess, 70, 73
Heymans, 175, 176, 177, 187
Hickling, 34, 84, 133
Hill, 28, 117, 185, 188
Hinomiya, 30
Hollaender, 222
Holm, 115
Homberg, 96, 99, 108
Hooke, 127
Howes, 93, 95
Hoyle, 76
Hudson, 66
Hughes, 202, 209
Hulme, 4, 6, 26, 52
Huntress, 118
Huygens, 88
Hykes, 187

Illig, 63
Imhof, 108
Inman, 29
Ishikawa, 78
Isaac, 67
Issatchenko, 67, 198
Ives, 101, 146, 199, 201, 202, 207, 209

Johann, 83
Johnson, 101, 126, 154, 157, 171, 172, 184, 185, 190, 191
Jordan, 199
Joubin, 76
Jourdan, 60
Jousset de Bellesme, 166
Julin, 81

Kaempfer, 72
Kalberer, 117
Kanda, 53, 124, 137, 139
Kastle, 171
Kato, 53
Kautsky, 115, 121
Kawamura, 41
Kayser, 94
Keenan, 218

Kemp, 63
Kiernik, 60
Kimler, 192
Kincaid, 220
King, 166
Kircher, 61, 103
Kishitani, 28, 33, 34, 36, 78
Klas, 188
Knab, 30, 66
Koch, 65
Kofoid, 43, 45, 46, 174
Kölliker, 68, 70, 166
Komarek, 56
Korr, 127, 144, 156, 177, 190, 192
Kortum, 99
Kratzenstein, 71
Krekel, 57
Kugelmass, 218
Kuhnt, 31
Kunkel, 99
Kutchera, 60
Kzevkin, 106

La Galla, 98
Ladd-Franklin, 17
Lamarck, 45
Langley, 199, 200, 206, 207
Langsdorff, 21
Lankester, 200
Lassar, 26
Le Roi, 44
Le Gentil, 44
Ledenfeld, 83
Lehman, 104
Leighton, 104
Lemon, 107
Lenard, 93, 107, 115
Lesueur, 81
Leuckart, 83
Levsin, 106
Lewis, 98
Leydig, 83
Licetus, 98
Lifschitz, 117
Linnemann, 114

INDEX

Linnaeus, 18
Lloyd, 104
Lode, 199
Loffling, 49
Lommel, 117
Ludwig, 26, 30, 40, 65, 66
Lund, 70, 166, 170

Macaire, 166
Macartney, 11, 45, 124
Macfayden, 28
Macrae, 40
Malisoff, 107
Maluf, 68, 168
Mangold, 79, 83, 85
Mark, 58
Martin, 26
Massart, 124, 162, 174
Mast, 73
Mathews, 218
Mathur, 121
Matuschek, 218
Maulik, 216
Mayen, 46
Mayer, 11
McAlpine, 40
McDermott, 73, 115, 135, 146, 171, 202
McKenny, 172, 183
McQuarrie, 218
Meissner, 30, 34
Mentzel, 108
Merritt, 94
Meyer, 60, 77
Michaelis, 6, 46, 54
Millikan, 154
Moeller, 117
Molish, 27, 39, 40, 54, 130, 183, 198, 201, 217
Monardes, 103
Monti, 75
Moore, A. R., 50, 110, 164, 175, 176, 177, 187
Moore, B., 174

Morley, 218
Mornay, 17
Morrison, 73, 124, 180
Mortara, 35, 76, 77
Moser, 217
Muraoka, 216, 217
Murray, 63, 69

Nadson, 198
Naef, 76
Nees von Esenbeck, 40
Neitzke, 115, 121
Nelson, 109
Newall, 98
Nenning, 218
Newton, 88-90, 93, 103
Nichols, 93, 94, 95, 100, 199
Niepce, 217, 218
Nightingale, 121, 199
Noggerath, 41
Nollet, 57
Nuesh, 26
Nunnemacher, 63

Okada, 36, 47, 52, 60, 62, 68, 70, 78
Oldenburg, 96, 99
Omeliansky, 180
Oshima, 83, 85
Osorio, 30, 84
Osten-Sacken, 73
Owsjannikow, 70

Panceri, 49, 50, 51, 75, 76, 80, 81, 130, 175, 176, 202
Papin, 9
Parker, A. S., 118, 214
Parker, G. H., 51, 164, 174
Parker, H. W., 16
Passerini, 65
Pasteur, 200
Paullinus, 26
Peron, 81, 202
Peters, 52, 164, 174, 177
Petrikaln, 114, 121

[315]

INDEX

Petsch, 119
Pfannsteil, 118
Pfeiffer, 29, 118
Pflüger, 6, 26, 130
Pfund, 94
Phipson, 18, 19, 20, 129
Picard, 105
Pickel, 111
Pickering, 69, 199
Pierantoni, 15, 30, 34, 35, 56, 77, 81
Pliny, 6, 16, 49, 75
Plot, 41
Polanyi, 113
Polatskii, 106
Polimanti, 81, 202
Pontus, 111
Pope, 109
Potts, 58
Pratje, 45, 162
Precht, 108
Priestley, 20, 21
Pringsheim, 93
Purkinje, 196

Quatrefages, 45, 160, 162

Radley, 104
Radziszewsky, 114, 121, 130
Ray, 65
Rayleigh, 114
Réaumur, 75, 122
Reboul, 114
Reichenspergers, 79
Reichert, 104
Reinhardt, 69, 83
Reinke, 46
Richard, 65
Richmond, 73
Richter, 183
Rigaud, 45
Risbec, 74
Risso, 82
Robin, 68

Root, 180
Rose, 111
Ruedemann, 166
Rupp, 101
Russel, 114, 217, 218

Sachs, 130
Sanzo, 83
Sars, 82
Schales, 118
Scharff, 114
Shapiro, 125
Schiller, 111
Schlaepfer, 218
Schmidlin, 117
Schmidt, 67
Schmitt, 93, 99, 112
Schmitz, 41, 117
Schönwald, 111
Schorigin, 111, 115
Schultes, 106
Schultze, 70
Schurig, 217
Schwarz, 112
Seaman, 68
Seidell, 114
Seitz, 101
Shima, 35
Shoji, 78
Shoup, 28, 125, 126, 192
Sidot, 99, 102
Silberschlag, 11
Singer, 104
Skoric, 188
Skowron, 35, 56, 77, 85, 87, 179, 189
Singh, 216
Slabber, 45, 49
Sloane, 68
Smith, 73
Snell, 152, 166, 167, 169
Snow, 95
Snyder, 166
Sokolow, 79

[316]

INDEX

Spallanzani, 40, 49, 50, 122, 166
Sparshall, 44
Spleist, 71
Sponer, 98
Stammer, 29, 67
Stanley, 118
Steche, 31, 83, 200
Sterzinger, 79
Stevens, 199
Stier, 80
Stokes, 101, 103, 146
Stradling, 223
Straubel, 99
Strutt, 98
Stubbes, 68
Stuchthey, 114, 220
Suchsland, 179, 217
Sugino, 36, 78
Sumner, 174
Suriray, 45
Svesnikov, 118
Swezy, 46

Tagaki, 36, 62, 78
Takase, 183, 202, 204
Tarchanoff, 14
Taylor, G. W., 171, 189, 191, 192
Taylor, H. S., 94, 113
Terao, 63, 64
Thiele, 115
Thielert, 118
Thomas, 214
Thomson, 50, 82
Tilesius, 46, 50, 52
Todd, 166
Tomaschek, 93
Townsend, 70
Trautz, 109, 111, 114, 115, 121, 154
Trojan, 47, 57, 63, 64, 76, 79, 83
Tschugaeff, 109
Tyndall, 103

Usow, 83

Vallentin, 63
van der Burg, 126, 157, 172, 190
van Musschenbrock, 108
van Schouwenburg, 121, 126, 157, 171, 172, 188-90, 202-4, 209, 214
Vanimo, 101
Vejdovsky, 59
Verany, 76
Very, 200, 206, 207
Verworn, 166
Vessichelli, 76
Vianelli, 57
Vignal, 162
Ville, 147
Vincent, 218
Viviani, 53, 79
Vogel, 70, 71
von Humboldt, 40
von Jhering, 69
von Prowazek, 104
Vormolen, 180
Vouk, 188

Wahlberg, 67
Watanabe, 62
Watase, 78, 160
Watermann, 63
Wedekind, 117
Wedgewood, 108
Weiser, 96, 107, 111, 112, 114, 121
Weiss, 113
Weitlaner, 115
Welch, 58, 59, 64, 104, 174
Weyenbert, 69
Wheeler, 30, 66
Wick, 97, 109
Wiedemann, 93, 112
Wielowiejski, 70
Wien, 99
Wilber, 93, 95
Will, 57
Williams, 66, 70
Winkelmann, 99
Witte, 118

INDEX

Wolf, 115
Wood, 102, 171, 223
Wulf, 220

Yasaki, 33
Yatsu, 52, 62
Young, 200

Zacharias, 46, 174
Zellner, 118
Zirpolo, 15, 31, 34, 180, 181, 183, 191
Zisch, 220
Zocher, 115
Zucchi, 98

SUBJECT INDEX

Abralia, 36
Abylopsis, 50
Acanthephyra, 63, 64, 133, 134, 161
Accelerators of respiration, 191-3
Acetic acid, 136
Acetone, 135, 137
Acholoë, 56, 146
Achorutes, 66
Achromobacter, 26, 184, 191, 192
Acropoma, 31, 84
Adrenalin, 85, 86-7, 165, 169, 191
Aequorea, 49, 127, 138, 175
Agaricus, 40
Aglaophenia, 48
Albumin, 136
Alciopinids, 47, 60
Alcohols, 135-8, 171, 172, 191
Alcyonarians, 50
Aldehyde, 114, 155
Alga, 13, 16, 27, 47
Algama, 50
Alkaloids, 109
Alkaptonuria, 21
Alpha methylglucoside, 192
Alpha naphthol, 136
Aluminum, 5, 107, 118, 142, 218
Amalopenaeus, 63
Amarin, 121, 145
Aminophthalichydrazid, 108, 117-18, 121, 145, 148, 214
Ammonium, 136, 137, 184, 188
Amphidinium, 45
Amphiura, 79, 175
Anaerobes, 29
Anaerobiosis, 126, 129
Anesthesia, 136, 171-3
Anisidin, 114
Annelida, 43, 55-60, 175
Anode, 5, 97, 107, 108, 118, 163
Anodoluminescence, 97

Anomalops, 14, 31-3, 84, 86, 87, 160
Ant, 30, 66
Antiluciferase, 140
Antipatharians, 50
Apatite, 96
Appendicularia, 80
Arctia, 67
Argon, 5
Argyropelecus, 84
Armillaria, 40, 41, 202
Arsenious acid, 111
Arsenite, 188, 190
Arthropoda, 43, 60-74
Ascidians, 12, 25, 80-2, 175
Ascomycetes, 40
Asparagin, 109, 115
Astraptor, 69, 200
Astronesthes, 83, 84
Azides, 220

Bacillus, 26, 172, 217
Bacteria, 6, 14, 15, 20, 21, 24, 25-36, 39, 40, 54, 61, 62, 65, 67, 76-7, 80, 81, 84, 86, 104, 122-7, 160, 170, 172, 175, 179, 180, 183, 188-90, 193, 198-204, 209, 212-14, 217
Balanoglossids, 25, 80, 123, 174, 175
Baldouin's phosphorus, 98
Balloelectric potentials, 106, 107
Barium, 4, 96, 98, 187
Beetles, 66, 67-74, 160, 161, 200
Benthobatis, 82
Benzene, 137, 138
Benzoic acid, 109
Benzoyl chloride, 137
Benzoyl peroxide, 145
Beroë, 52, 177
Bichromates, 145

[319]

INDEX

Bile salts, 185
Birds, 15, 17
Black body, 91, 92
Blepharocysta, 45
Blood, 218
Boletophila, 66, 67, 161
Bolinopsis, 52
Bolognian phosphorus, 3, 98, 108
Bolometer, 206, 207
Bone, 102
Botryllus, 81
Branchiopoda, 42, 55
Brisingia, 15
Brittle stars, 12, 13, 25, 79-80, 123
Bromide, 107, 183, 184, 187
Bryozoa, 42, 54
Bupestris, 68
Butterflies, 15

Calcite, 110
Calcium, 4, 96, 98, 99, 108, 171, 183-7
Campanularia, 48
Canal rays, 99
Cancer, 61
Candle-fly, 13
Candoluminescence, 94, 95
Canton's phosphorus, 4, 99
Cadmium, 142
Carbon dioxide, 127-9, 130, 212
Carbon disulphide, 137
Carbon monoxide, 189, 190
Carbon tetrachloride, 135, 137
Cartilage, 102
Catalase, 145
Caterpillars, 14, 29
Cathode, 107, 163
Cathode rays, 4, 97, 99, 102, 109, 110, 112
Cathodoluminescence, 97
Cavernularia, 50, 175, 189
Cavitation, 5, 106, 107
Celluloid, 110
Centrosyllium, 83
Cephalopods, 12, 23, 76-9

Ceratias, 34, 84
Ceratium, 45, 174
Ceratoplanus, 67
Cerebrin, 114
Cestus, 52, 177
Chaetognatha, 43, 55
Chaetopterid, 57
Chaetopterus, 12, 56, 57, 75, 161, 175, 200
Charybdea, 49
Chauliodus, 84, 85, 87
Cheese, 41
Chemiluminescence, 23, 88, 113-21, 130, 177, 203, 212, 214, 219, 220
Chilota, 55
Chiridius, 62
Chiroteuthis, 36
Chitin, 110, 207
Chloretone, 172
Chlorides, 111, 112, 131, 183-7
Chlorine, 220
Chloroform, 133, 135-7, 168, 171, 172
Chlorophane, 96
Chlorophyll, 27, 103, 116, 198
Chlorpicrin, 117
Cholesterin, 114
Cholic acid, 114
Chordata, 43, 86-7
Chromophyton, 16
Chrysaora, 49
Ciona, 35, 81
Cirratulid, 56, 57
Civit cat, 21
Clam, 75
Clavula, 49
Cleodora (*Clio*), 75
Clitocybe, 40, 41
Clytia, 48
Coal, 37-9
Coccobacillus, 26, 31, 35
Cochlodinium, 45
Coelenterata, 42, 48-52, 164
Coelorhynchus, 34, 84, 202, 204
Co-enzyme, 132

[320]

INDEX

Collosendeis, 65
Collosphaera, 46
Collozoum, 46, 175
Collybia, 40
Color of light, 74, 79, 97, 109, 196, 200-4, 219
Co-luciferase, 132, 141
Comb jellies, 7, 52-3, 177
Conchoëcia, 61
Copepods, 15, 62, 122, 123, 174, 175
Copper, 95, 98, 99, 145
Corallines, 25
Corposants, 18
Corticum, 40
Corycaeus, 62
Corynocephalus, 56
Crab, 61
Crateromorpha, 47
Creseis, 75
Crustaceans, 11, 13, 14, 23, 25, 60, 61-4, 131
Crystalloluminescence, 94, 111-12
Ctenophores, 12, 25, 31, 42, 52-3, 127, 161, 176, 186, 199
Culture medium, 27, 33
Cyanea, 49
Cyanide, 126, 156, 171, 188-90, 213
Cyanosin, 139
Cyclops, 61
Cyclothone, 85
Cypridina, 61, 122, 124, 125, 128, 131, 133, 134, 136-42, 145, 146, 149, 154-7, 161, 170, 172, 175, 179, 187, 189, 198, 199, 202, 203, 209, 214
Cytolysis, 49, 50, 132, 185

Decapods, 62-4
Dehydrogenation, 142-4, 155-7, 159
Deuterium oxide, 153, 191
Diadema, 15
Diamond, 3, 96, 102, 108
Dictyophora, 40

Dimethyldiacrylidium, 119, 121, 145, 203
Dioxymethylen, 114
Diphyes, 50
Doliolum, 80
Drying, 41, 61, 122-3

Earthworms, 12, 25, 31, 55-6, 65, 123, 179, 189
Echinodermata, 43, 79-80
Echiostoma, 86, 87, 165
Efficiency, 88, 204-16
Eggs, 13, 26, 45, 52, 71
Egg yolk, 100
Eisenia, 55, 56
Elaterid beetle, 68-9
Electra, 54
Electrical discharges, 18, 20, 109
Electroluminescence, 5, 102, 105-6, 216
Electrolytic flames, 107
Eledonella, 76
Embryos, 13, 52
Emplectonema, 53
Enchytraeus, 55
Enoproteuthis, 36
Eosin, 116, 139
Equula, 34, 84
Esculin, 104, 114, 145, 147, 148
Ether, 132, 137, 169, 171, 172
Etmopterus, 83
Euchaeta, 62
Eucharis, 52
Eudia, 67
Euphasia, 63
Euphorbia, 17
Euprvmna, 35, 76
Eusyllis, 56
Eutryphoeus, 55, 56
Evolution of luminescence, 23, 158-9
Excess luminescence, 126
Excited atoms, 88, 89, 95, 102, 105, 115
Eye, 16, 23, 63, 77, 85, 174, 178, 195-6, 204, 210

[321]

INDEX

Fatigue, 53, 163
Ferricyanide, 117-19, 144, 145
Filters, 28
Fireflies, 5, 12, 13, 15, 31, 67-74, 112, 122-4, 131, 133, 161, 171, 173, 175, 178, 189, 199-202, 204, 207, 209, 211, 216
Fish, 12, 14, 15, 23, 25, 26, 30-4, 82-7, 160, 161, 165, 175, 189, 200
Flagellates, 12, 25, 43-6, 48, 54, 109, 129, 161, 174
Flash, 126, 127, 161-72, 178, 179
Flashing of flowers, 18
Flavianic acid, 137
Flavin, 138, 139, 193, 203
Flesh, 6, 20, 26, 123, 127
Flies, 66-7
Fluorescence, 4, 16, 97, 102-5, 113, 116, 118, 137, 177, 187, 198, 199, 216, 218
Fluorescence microscope, 104
Fluoride, 187, 188, 190
Fluorspar, 96
Flustra, 54
Fomes, 40
Fontaria, 65
Food and luminescence, 126, 191
Formaldehyde, 114, 115
Fovea, 5, 196
Fox-fire, 36
Frogs, 14
Fucus, 27
Fulgora, 15
Fungi, 7, 17, 20, 24, 25, 36-41, 65, 122, 123, 175, 202
Funiculina, 50
Fur, 105

Gallic acid, 114, 148
Galvanoluminescence, 94, 107-8
Gamma-rays, 97
Gasteromycetes, 40
Gastrosacchus, 63
Gennadas, 63

Geophilus, 65
Gigantocypris, 61
Glow-worms, 5, 6, 12, 15, 68, 199
Glucose, 114, 126, 127, 184, 191, 220
Glue, 100, 115
Glycerine, 27, 106, 114, 136, 191, 213
Glycocholic acid, 114
Gnathophausia, 63
Gonostoma, 85
Gonothyrea, 48
Gonyaulax, 45, 46
Gorgonians, 50
Grantia, 47
Granules, 15, 24, 28, 36, 45, 47, 49, 50, 56, 62, 65, 69, 75, 84, 112, 114, 131, 132-3, 161, 187, 189
Grignard compounds, 108, 117, 214
Guaiac gum, 136
Gymnodinium, 45

Halistaura, 49
Heat production, 128
Heliotropism, 198
Helix, 75
Hematine, 118, 135
Hematoporyphrin, 104
Hemin, 157
Hemoglobin, 104, 118, 143, 145, 147, 148
Henlea, 55
Heredity of luminous fungi, 40
Heterocarpus, 63
Heterocirrus, 56
Heterorhabus, 62
Heteroteuthis, 35, 77, 134, 161, 179, 189
Himantarium, 65
H ion concentration, 27, 153, 184, 187-8, 202
Hippopodius, 50
Hippuric acid, 109
Histioteuthis, 76
Homberg's phosphorus, 99, 108

[322]

INDEX

Homogentistic acid, 21
Hormones, 84, 85, 86, 191
Humus, 115
Hyalea (Cavolina), 75
Hydroanisamid, 114
Hydrobenzamid, 114
Hydrocinamid, 114
Hydrocuminamid, 114
Hydrogen, 118, 124, 142
Hydrogen peroxide, 115-19, 131, 135, 145, 147, 218
Hydroids, 12, 25, 48-9
Hydrosulphite, 127, 142
Hymenocephalus, 34
Hypobromites, 145
Hypochlorite, 117, 119, 135, 145
Hypoiodites, 145

Ice, 108, 111
Ignis fatuus, 19, 20, 93
Ileodictyon, 40
Illumination, 177-9, 196-8
Incandescence, 90-4
Indigo, 192
Indophenol, 144
Infection with luminous bacteria, 6, 14, 20, 26-36, 66, 67, 76, 77, 84
Infrared rays, 101, 207
Inhibition by light, 173-9
Inhibition of respiration, 188-92
Insects, 25, 42, 60, 66-74
Intensity of light, 125, 126, 196-200
Invisible radiation, 216-20
Iodide, 183, 184, 187
Iodoacetate, 188, 190
Ionization, 89
Iridescence, 15, 16
Iron, 136, 145
Isistius, 83

Jelly-fish, 7, 12, 25, 48-50, 186

Kalchbrennera, 40
Kinetics of luminescence, 148-157
Kingcrab, 60

Lactic acid, 168
Lampyris, 71
Lantern-fly, 13
Laodicea, 49
Larvae, 13, 62, 67, 79, 161, 199
Latex, 17
Lead, 98, 140
Leaves, 41
Lecithin, 114, 219
Leioptilus (Ptylosarcus), 50
Lens of eye, 102
Light and organisms, 39, 63, 81, 166, 173-80
Light, nature of, 88-94, 194-8
Light units, 196-8
Lignum nephriticum, 103
Lipura, 66
Liquid air, 28, 181
Linophryne, 87
Lithium, 98, 112, 184, 187
Lizard, 16
Locellina, 40
Loligo, 35, 76
Lophin, 104, 114, 145, 147
Lucicutia, 62
Luciferase, 24, 28, 34, 54, 62, 75, 126, 130-58, 161, 170-2, 187, 189, 221
Luciferesceine, 145-6
Luciferin, 24, 28, 34, 54, 62, 75, 126-8, 130-58, 161, 170-2, 179, 187, 189, 209, 221
Lucigenines, 119
Luciola, 131, 166, 168, 175, 178
Luminescence distribution, 12
Luminol, 108, 117-18, 121, 145, 148, 214
Luminous flux, 197
Lycoteuthis, 76, 200
Lyoluminescence, 94, 112

Macrochaeta, 56
Magnesium, 96, 137, 142, 171, 183-7

INDEX

Malacocephalus, 30, 34, 84, 133, 138, 189
Malic acid, 109
Man, 20, 21, 26
Manganese, 96, 98, 142, 145
Mannite, 109, 114
Marasmius, 40
Marble, 96
Mating signal, 58, 59, 73, 74
Maurolicus, 84, 85
Mayflies, 14, 30
Mechanical equivalent of light, 198
Medusae, 31, 45, 48-50, 53, 122, 127, 161, 175
Meganyctiphanes, 63
Megascolex, 55, 56
Melanoteuthis, 76
Membranipora, 54
Mercury, 5, 105, 107, 207, 220
Mesochaetopterus, 57
Methane, 19, 20
Methylchlorophyllid, 116
Methylene blue, 139, 143, 144, 189, 192
Metridia, 62
Mica, 3
Micrococcus, 26, 34, 184
Microscolex, 55, 56, 175
Microspira, 26, 30
Midges, 14, 67
Millipeds, 65
Mitogenetic rays, 221-3
Mitotase, 221
Mitotin, 221
Mitrocoma, 138, 175
Mnemiopsis, 52, 164, 177, 199
Models of bioluminescence, 147-8
Mole crickets, 14, 30
Molluscs, 12, 25, 43, 74-9
Monocentris, 33, 84, 86, 175
Moss, 16
Moths, 16, 66, 67
Mutants, 182, 193
Mycelium X, 39
Mycena, 40, 41

Myctophum, 84, 85, 87
Myriapods, 12, 25, 60, 65, 112, 122, 161
Mysis, 63
Mytilus, 132
Myxosphaera, 46

Narcosis, 156, 171-3
Neanura, 42, 66
Necco wafers, 110
Nemathelminthes, 25, 42, 53
Nematocelis, 63
Nematolampas, 76, 79
Nemertea, 42, 53-4
Neon, 5, 105, 106, 110, 216
Nereids, 56
Nerves, 17, 71, 75, 82, 123, 161, 165, 166, 168, 170, 173, 176, 178, 187, 223
Nitrate, 187
Nitrobenzol, 106
Nitrogen, 5, 98, 110, 119
Nitrophenols, 192, 193
Noctiluca, 7, 11, 44-5, 57, 98, 123, 124, 133, 160, 162-4, 167, 172-5, 182, 185-9
Noctilucine, 129
N rays, 223
Nucleoprotein, 136
Nudibranch, 75, 76
Nyctiphanes, 63, 64

Obelia, 48
Octochaetus, 55
Octopods, 76
Ocyropsis, 52
Odontosyllis, 56-9, 133
Oil drops, 46, 50, 52, 114, 130, 145, 218
Omphalia, 40, 41
Onchaea, 62
Oniscus, 61
Onychiurus, 66
Onychophores, 60
Ophiacantha, 79

[324]

INDEX

Ophionereis, 79
Ophiopsila, 79
Ophioscolex, 79
Ophiothrix, 79
Oplophorus, 63
Orphnaeus, 65
Orya, 65
Osmium, 119
Ostracods, 61-2, 122, 123, 131, 175
Oxidases, 145
Oxygen, 27-9, 84, 108, 110, 113, 114, 117-19, 123-30, 147, 148, 154, 157, 162, 168-73, 176, 177, 189, 212-14, 220
Oxygen consumption, 125, 126, 179, 181, 184, 188, 189, 212-14
Oxyluciferin, 141-5
Oxyluminescence, 113
Ozone, 3, 108, 114, 117, 119, 220

Palladium 142
Pandanus, 18
Panus, 40, 175
Papaverin, 114
Paraphenylenediamine, 136
Parascinia, 67
Paussus, 68
Peat, 41
Pelagia, 49, 50, 51, 122, 127, 146, 175, 187
Pennatulids, 50-1, 53, 122, 123, 161, 174, 175, 189
Peptone, 130, 191, 213
Perborates, 145
Perchlorates, 145
Peridinium, 45, 46
Permanganate, 115, 131, 135, 145, 220
Peroxidases, 118, 145, 147, 148
Persulphates, 145
Phaeoporphyrin, 116
Phengodes, 24, 69, 160, 200
Phenol, 114, 155
Phialidium, 49, 175
Pholadidea, 132

Pholas, 12, 35, 75, 122, 123, 131-3, 135, 136, 141, 146, 161, 179
Phoronidea, 42
Phosphaenus, 167
Phosphate, 98, 153, 184, 185, 188
Phosphine, 19
Phospholipin, 139
Phosphorescence, 4, 97-101, 113, 198, 207, 219, 223
Phosphoroscope, 100
Phosphors, 3, 4, 98, 105, 108, 130
Phosphorus, 3, 44, 90, 96-9, 113-14, 118, 121, 129, 130, 214
Phosphotungstic acid, 137, 140
Photinus, 74, 131, 134, 178, 199-201, 207, 209
Photobacterium, 26, 126, 202, 204
Photoblepharon, 14, 31-4, 84, 86, 87, 160, 189, 200
Photochemical reaction, 23, 88, 177
Photogen, 129-30
Photogenin, 132
Photographic plate, 115, 198, 217, 219, 222
Photoluminescence, 97
Photophelein, 132-3
Photophores, 63-4, 78-9, 83-5, 161
Photoplasma, 130
Photosynthesis, 27
Photuris, 71, 131, 134, 165, 169, 178, 199, 201, 209
Phthalocyanin, 116
Phyllirrhoë, 12, 75, 76, 122, 134
Physiculus, 31, 34, 84
Physodera, 68
Picric acid, 136, 137, 140
Piezoluminescence, 94, 108-10
Planck's law, 91
Platinocyanide, 97, 102, 218
Platinum, 142
Platyhelminthes, 25, 42, 53
Plesiopenaeus, 63
Pleurobrachia, 52
Pleuromma, 62

[325]

INDEX

Pleurotus, 40, 41
Plocamopherus, 75
Polarized light, 103, 194, 198
Polycheles, 63
Polycirrus, 57
Polyhydroxybenzene, 140
Polynoëids, 56, 60
Polyporus, 40
Pontella, 62
Porichthys, 85-6, 165
Porifera, 42, 47-8
Potassium, 111, 112, 114, 183-7
Potatoes, 41
Praya, 50
Pressure wave, 166, 167, 170
Proctoporus, 16
Proluciferin (preluciferin), 140-1
Prorocentrum, 45
Proteose, 136, 139
Protozoa, 23, 42-7, 161, 162
Pseudoflash, 170
Pseudomonas, 26, 203
Pteroides, 174, 176
Ptychodera, 80, 174, 175
Ptylosarcus, 175
Purkinje effect, 196
Purple arcs, 17
Pycnogonid, 65
Pyrocypris, 61, 133, 134
Pyrocystis, 45
Pyrodinium, 45
Pyrogallol, 114, 115, 121, 136, 145, 148, 220
Pyroluminescence, 94, 95, 113
Pyrophorine, 146
Pyrophorus, 68-9, 130, 135, 138, 145, 146, 161, 167, 178, 198-201, 206
Pyrophosphate, 188, 190
Pyrosoma, 12, 31, 81-2, 122, 134, 138, 175, 176, 202

Quanta, 88, 214, 219, 222
Quartz, 3, 96, 108
Quinine, 103, 187
Quinone, 144

Radiant flux, 91, 197
Radiolaria, 25, 46-7, 127, 161, 175
Radium, 5, 99, 109, 180, 218
Raindrops, 18, 19
Rathkea, 49
Rectifiers, 5
Redox potential, 140, 144, 192
Reduction of oxidized luciferin, 141-5
Reflection, 16, 17
Renilla, 50, 51, 164, 174
Resonance radiation, 102
Retepora, 54
Retina, 5, 17, 195-6
Rhizomorpha, 7, 40, 41
Rhizostoma, 49
Rhodamine, 104, 116
Rhythms of luminescence, 173-4, 178
Rondeletia, 34, 76
Rotifers, 54
Rubber, 3, 105
Rubies, 102
Russel effect, 217-18

Saccharin, 109
Sagitta, 55
Salacylamid, 110
Salacylic acid, 100
Salophen, 110
Salpa, 31, 80
Salts, effect of, 153, 183-7, 202
Sand fleas, 14, 29
Santonin, 109
Saponin, 132, 185
Sappharina, 15
Sarcina, 6
Scheelite, 102
Schistostega, 16
Schizopods, 62-4
Scintillations, 110
Scolioplanes, 65
Scolopendra, 57

INDEX

Scrupocellaria, 54
Sea fans, 50
Sea pansy, 51
Sea pens, 12, 25, 50-1
Sea phosphorescence, 7-11, 43-6
Sea urchin, 15
Selachians, 82-4
Sensitization, 102, 104
Sepia, 35
Sepiola, 34, 35, 76
Sergestes, 63, 64
Sertularia, 48
Sex attraction, 58-9, 73, 74, 79, 87, 165
Sharks, 82-4
Shrimp, 14, 30, 161, 165
Sidot blend, 99, 102
Siloxene, 115-16, 121
Silk, 105
Silver, 98
Siphonophores, 25, 50
Siriella, 63
Siriope, 49
Skin, 20
Snowflakes, 19
Sodium, 5, 111-14, 131, 183-9
Solinissus, 49
Somniosus, 83
Sonic chemiluminescence, 107
Sonoluminescence, 5, 106-7
Specificity, 133-4
Spectra, 89, 100, 109, 110, 112, 113, 116, 119, 121, 200-4, 220-2
Spectral energy curves, 92, 121, 194
Spectral sensitivity curves, 195
Sphaerozoum, 46, 175
Spiders, 25, 30, 61, 64-5, 171
Spinthariscope, 5, 171
Sponges, 25, 47-8
Springtails, 66
Squid, 12, 25, 31, 34-6, 161, 165, 175, 189, 200
St. Elmo's fire, 18
Starch, 100
Star-fish, 15

Stefan-Bolzman law, 91
Stigmatogaster, 65
Stimulation, 24, 45, 46, 49, 51, 53, 76, 81, 85, 160-72, 180, 182
Stokes law, 101
Stomatoca, 49, 175
Stomias, 84
Strontium, 96, 98, 187
Styliola, 75
Stylocheiron, 63
Sugar, 3, 108, 109, 110, 112, 136, 202, 212
Sulphates, 96, 98, 111, 187
Sulphides, 4, 5, 44, 98, 99, 101, 102, 119, 142, 207, 218, 219
Sulphocyanide, 187
Supersound waves, 106
Sweat, 20, 21
Symbiosis, 14, 15, 30-6, 56, 67, 80, 81, 160, 202, 204
Synchaeta, 54
Synchronous flashing, 72-4
Systellaspis, 63

Tannic acid, 114, 137, 140
Tape, 3, 105-6
Tartaric acid, 109
Taurocholic acid, 114
Teeth, 102
Teleosts, 82-7
Temperature, 90-2, 95, 96, 99, 100, 128, 166, 171, 177, 180-2, 202, 207
Tendon, 102
Terebellids, 56, 57
Termites, 30
Thalassicola, 46, 175
Thalassocaris, 63
Thelepus, 57, 146
Thermoluminescence, 94, 95-6
Thiophosgene, 115
Thymol, 172
Thysanoëssa, 63
Thysanopoda, 63
Toluol, 135, 137